Compressed Sensing in LiFi and WiFi Networks

To my mother, with my profound love and thanks.

Compressed Sensing in LiFi and WiFi Networks

Malek Benslama
Hatem Mokhtari

ISTE PRESS

ELSEVIER

First published 2017 in Great Britain and the United States by ISTE Press Ltd and Elsevier Ltd

ISTE Press Ltd
27-37 St George's Road
London SW19 4EU
UK

www.iste.co.uk

Elsevier Ltd
The Boulevard, Langford Lane
Kidlington, Oxford, OX5 1GB
UK

www.elsevier.com

For information on all our publications visit our website at http://store.elsevier.com/

British Library Cataloguing-in-Publication Data
A CIP record for this book is available from the British Library
Library of Congress Cataloging in Publication Data
A catalog record for this book is available from the Library of Congress
ISBN 978-1-78548-200-7

Contents

Preface

This book follows four other books published by ISTE. I would like to express my admiration and gratitude to this publishing house, which brings a decisive contribution to publishing and distributing books by French-speaking scientific authors.

I could not forget Professor Guy Pujolle, editor of the Network and Telecommunications series in which this book is published; I am infinitely grateful to him.

The idea of a book on compressed sensing is the result of discussions that I have had with my colleague Hadj Batatia, who helped with a presentation on compressed sensing by Yonina Eldar during his stay in Toulouse. I admit I was intrigued by this new, innovative theory that opens up new perspectives for research on signal processing and that puts the whole framework of Shannon's theory into question. Starting with a literature review on compressed sensing, I discovered that interesting applications on compressed sensing in quantum intrication were possible. In collaboration with Hadj Batatia on the compressed sensing aspect and Hatem Mokhtari on the telecommunications network aspect, I began to work on this book by examining the possible applications for compressed sensing in LiFi and WiFi systems. I also undertook the integration of new developments in compressed sensing and radar tracking.

Malek BENSLAMA
Hatem MOKHTARI
September 2017

List of Acronyms

16APSK	16 Asymetric Phase Shift keying
32APSK	32 Asymetric Phase Shift keying
3GPP	Third Generation Partnership Project
8APSK	8 Asymetric Phase Shift keying
AAA	Authentication Authorization Accounting
AAL5	ATM Adaptation Layer 5
ACK	Acknowledge
ACQ	Acquisition
AES-CCM	Advanced Encryption Standard CCM (Counter with CBC-MAC) mode
AJ	Adjustment
ANDSF	Access Network Discovery and Selection Function
AP	Application Protocol
ARIB	Association of Radio Industries and Businesses
AS	Autonomous System
ASCONF-ACK	Address Configuration Acknowledgment
ASN	Access Service Network

ASNGW	Access Service Network Gateway
ATM	Asynchronous Transfer Mode
AuC	Authentication Center
BC	Box Counting
BCH	Bose Chaudhuri Hocquenghem
BER	Bit Error Rate
BGP	Border Gateway Protocol
BS	BackSpace
BTS	Base Transceiver Station
CAC	Call Admission Control
CDMA	Code Division Multiple Access
CMAC	Cipher-MAC
CN	Core Network
CNES	*Centre National d'Etudes Spatiales*
CRC	Cyclic Redundancy Check
CS	Command Service
CSC	Common Signaling Channel
CSN	Connectivity Service Network
CSTP	Control Stream Transmission Protocol
CWR	Congestion Window Reduced
D2-MAC	Duobinary Digital-Multiplex Analog Components
DAMA	Demand Assignment Multiple Access
DAR	Dynamic Address Reconfiguration
DCT	Discrete Cosinus Transform
DS	Digital Subscriber
DS/CDMA	Direct Sequence Code Division Multiple Access
DSL	Digital Subscriber Line

DVB	Digital Video Broadcast
DVB-C	Digital Video Broadcast over cable
DVB-CI	DVB Common Interface
DVB-CSA	DVB Common Scrambling Algorithm
DVB-RCS	Digital Video Broadcasting – Return Channel via Satellite
DVB-S	Digital Video Broadcast over Satellite
DVB-S/RCS	Digital Video Broadcast satellite Return Channel via Satellite
DVB-SI	Digital Video Broadcast-System Information
DVB-T	Digital Video Broadcast Terrestre
EAP	Extensible Authentification Protocol
ECNE	Explicit Congestion Notification Echo
EIR	Equipment Identity Register
EPC	Evolved Packet Core
EPG	Electronic Program Guides
ERS	European Remote Sensing
ES	Event Service
ES	Elementary Stream
ETSI	European Telecommunications Standards Institute
FBSS	Fast Base Station Switching
FDD	Frequency Division Duplex
FEC	Forward Error Control
FP	Frame Protocol
GEO	Geostationary
GGSN	Gateway GPRS Support Node
GMSC	Gateway MSC

GPRS	General Packet Radio Service
GSM	Global System for Mobile Communications
H-ARQ	Hybrid Automatic Repeat Request
HDTV	High Definition Television
HEO	High Earth Orbit
HLR	Home Location Register
HOL	Head of Line Blocking
HSDPA	High Speed Downlink Packet Access
HSPUS	High Speed Packet Uplink Stream
IAM	Interference Access Multiple
IANA	Internet Assigned Numbers Authority
IDU	Indoor Unit
IEEE	Institute of Electrical and Electronics Engineers
IETF	Internet Engineering Task Force
IID	Identical Independent distributed
IMS	IP Multimedia Subsystem
IP	Internet Protocol
IPoS	IP over Satellite
IS	Information Service
ISMP	Inter-System Mobility Policy
ISO	International Organization for Standardization
ISRP	Inter-System Routing Policy
IWU	Interworking Unit
JPL	Jet Propulsion laboratory
LDPC	Low-Density Parity Check
LENS	Low Error Noise Sparse
LEO	Low Earth Orbit

LOS	Line of Sight
LTE	Long-Term Evolution
MAC	Medium Access Control
MAN	Metropolitan Area Network
MANET	Mobile Adhoc Network
MDHO	Macro Diversity Handover
ME	Mobile Equipment
MEO	Major Equipment Operator
MEO2	Medium Earth Orbit
MF-TDMA	Multifrequency Time Division Multiple Access
MIH	Media Independent Handoff
MIHF	Media Independent Handover Function
MIMO	Multiple Input/Multiple Output
MME	Mobility Management Entity
MPE	Multiple Protocol Encapsulation
MPEG2	Moving Picture Experts Group 2
MPEG2-TS	Moving Picture Experts Group 2 Transport Stream
MPEG-TS	Moving Picture Expert Group Transport Stream
MS	Mobile Stream
MSC	Mobile Switching Center
MSC/VLR	Mobile Service Switching/Visitor Location Register
MSCTP	Mobile Stream Control Transmission Protocol
NAP	Network Access Provider
NCC	Network Control Center
NLOS	Non Line of Sight
NMMC	Network and Mission Management Center
NSP	Network Service Provider

ODU	Outdoor Unit
OFDM	Orthogonal Frequency Division Multiplexing
OFDMA	Orthogonal Frequency Division Multiple Access
OSI	Open Systems Interconnection
PAL	Phase Alternate Line
PDF	Probability Density function
PDN	Packet Data Gateway
PDU	Protocol Data Unit
PDV	Prise De Vue
PES	Packetized Elementary Stream
PF	Path Following
PID	Packet Identifier
PMTU	Path Maximum Transmission Unit
PN	Pseudo Noise
PS	Packet Switching Network
QG	Quantile Graphs
QPSK	Quadrature Phase Shift Keying
RAN	Radio Access Network
RCST	Return Channel Satellite Terminal
RNC	Radio Network Controller
RNS	Radio Network Subsystem
SAE	System Architecture Evolution
SAPs	Service Access Points
SAR	Synthetic Aperture Radar
SCTP	Stream Control Transmission Protocol
SECAM	Sequential Color and Memory

SGSN	Serving GPRS Support Node
SI	Simple Integration
SIGTRAN	Special Interest Group on Translations
SIM	Subscriber Identity Module
SIP	Session Initiation Protocol
SIR-B	Shuttle Imaging Radar-B
SLA	Service Level Agreement
SNDU	Sub Network Data Unit
SWR	Standing Wave Ratio
SWU	Separative Work Unit
SYNC	Synchronization
TCP	Transmission Control Protocol
TDD	Time Division Duplex
TIA	Telecommunications Industry Association
TRF	Traffic
TTA	Telecommunication Technology Association
TTC	Telecommunication Technology Committee
UDP	User Datagram Protocol
UE	User Equipment
UICC	Universal Integrated Circuit Card
ULE	Unidirectional Lightweight Encapsulation
UMTS	Universal Mobile Telecommunication System
URL	Uniform Resource Locator
USIM	Universal Subscriber Identity Module
UTRA	Universal Terrestrial Radio Access
UTRAN	UMTS Terrestrial Radio Access Network
VGSN	Virtual GPRS Support Node

VLR	Visitor Location Register
VoIP	Voice over IP
VSAT	Very Small Aperture Terminals
WAG	WiMAX Access Gateway
WCDMA	Wideband Code division Multiple Access
WiMAX	Worldwide Interoperability for Microwave Access
WLAN	Wireless Local Area Network
WMAN	Wireless Metropolitan Area Network
WPDG	Wimax Protocol Data Gateway

Introduction

In signal processing, the use of Shannon's theorems is commonplace. Serious processing without reference to the sampling theorem is not be imaginable. This processing is only possible if the signal meets the necessary conditions. The first attempt consisted of using the theory of wavelets to solve some of its aspects. Donoho and Mallat's research sometimes refers to this domain. However, some problems posed by parsimonious or sparse signals remain. Compressed sensing relies on the fact that most analog signals have a structure that can be used to reconstruct them. Knowing this structure makes it possible to reconstruct the signal using a sampling rate that is significantly lower than the Nyquist rate.

The difficulty lies in determining the structure, which is a difficult non-determinist polynomial (NP) problem that cannot generally be solved in a reasonable time span. But it is true that with a small mathematical trick even this is not necessary and the signal can in fact be reconstructed successfully with a fraction of Nyquist's sampling frequency. This leads us onto the theory of computational complexity. Consequently, it is a question of finding a polynomial algorithm to solve a NP-hardness problem.

Compressed sensing opens up a new revolution in processing problems covering mathematics, applied math, IT, information theory, signal processing, circuit design, biomedical imaging, electromagnetism and radar, tracking, etc. One of the principle contributors is Baraniuk [BAR 07, BAR 10].

Compressed sensing is a contemporary paradox, in fact we are in the presence of a massive acquisition of data, of which most is redundant and can therefore be ignored, which causes colossal amounts of waste.

The first contributions to the theory of compressed sensing are owed to Candés *et al.* [CAN 05, CAN 06a, CAN 06b, DON 06a]. Compressed sensing makes it possible to obtain signals at good resolution from a few sensors; acquisition is not adaptive and no effort at understanding the signal is needed. It is a simple process of acquisition followed by digital optimization.

Among the authorities on compressed sensing, we list Donoho [DON 06b] who has developed extensive research from the compression transform. This author, recognized as a fundamental point of reference in wavelet theory, tackles compressed sensing from the optimal signal reconstruction and information complexity; all aspects of compressed sensing are tackled: they include integrated procurement processing, optimal reconstruction, data complexity, adapted sampling, and decomposition following a minimal norm.

The restricted isometry property developed by Candes and colleagues [CAN 08a, CAN 08b] and its implications in compressed sensing form a major element. Another important element that ensures the transition of wavelets to compressed sensing is the iterative hard threshold developed by Blumensath and Davies [BLU 09].

The sparseness that is implicit in magnetic resonance (MR) images is used so as to significantly subsample k-spaces. Some MR images such as angiograms are already sparse in pixel representation [LUS 07]. Projection by gradient for sparse reconstruction is one application for compressed sensing. Many signal processing and statistical inference problems involve sparse solutions for underdetermined or badly conditioned equation systems. A standard approach consists of minimizing an objective function, which includes a quadratic function. This problem has also been studied by Figueiredo and co-authors; she is also interested in inverse problems and quadratic programming [FIG 07]. Applications with compressed sensing by minimizing norm l_1 from iterative Bregman algorithms are given by Yang and Zhang [YAN 11]. These authors suggest simple and extremely effective methods for solving the problem of basic tracking, e.g. $\min\{||u||_1 : Au = f, u \in \mathbb{R}^n\}$, which is widely used in compressed sensing. These methods are based on Bregmann iterative regularization and they give an exact solution after only a few sequences of calculation for the problem without constraints of $\min_{u \in \mathbb{R}^n} ||u||_1 + 1/2||Au-f^k||_2^2$ for a given matrix A and a vector f^k.

Tian and Giannakis have found in the paradigm from access to the open spectrum that radios dynamically detect the radioelectric spectrum's environment and they have to adjust their transmission parameters quickly to use the available spectrum. The envisaged radio agility requires a fast and precise spectrum on a large pass band, which calls into question the methods for spectral estimation that function using the Nyquist rate. The weak density of the spectral signal in free access networks calls for the development of compressed sensing techniques adapted for identifying the spectral hole. Nyquist subsamples are used to detect and classify frequency bands using wavelets. Because estimating the spectrum's location takes priority over reconstructing the signal to scale, new sensing algorithms are suggested, which have robustness to noise and can reduce the sampling rate [TIA 07].

Hermann and Strohmer studied high-resolution radar using compressed sensing. They suggested a radar styled by compressed sensing in which the frequency time plan is discretized into a square grid $N \times N$. Assuming that the number of K targets is low, i.e. $K \ll N^2$, they can transmit a sufficiently incoherent pulse and they use compressed sensing techniques to reconstruct the target scene. A theoretically sparse upper bound K is shown. Digital simulations confirm that better performances can be obtained in practice. This new compressed sensing approach offers great potential for a better radar resolution compared to classic radar [HER 09].

Rauhut and Schnass extend the concept of compressed sensing to signals that are not sparse. In an orthonormal base in the form of a redundant dictionary, they have demonstrated a matrix that is a composite of a determined set random matrix, which is a determinist dictionary that has small restricted isometric constants. Therefore, signals that are sparse according to the dictionary can be reconstructed via basic tracking from a small measurement of random measurements. Moreover, the threshold is studied as an algorithm for compressed sensing with conditions to guarantee reconstruction with a high probability. These different schemas are compared to digital experiments [RAU 08].

Compressed sensing makes it possible to carry out compression and detection based on the existence of sparse representation of the processed signal and a set of projections of the measurements.

Classically, compressed sensing generally assumes that the projections are taken at random. Elad considers how to optimize these projections. This optimization is directly prohibitive; it targets an average measurement of the mutual coherence of the effective dictionary and shows that this results in a better reconstruction. Basic tracking and adapted orthogonal tracking lead to new projections. The projections designed show that tracking is advantageous, with a reduction in the error rate by a factor of 10 or more [ELA 07].

A small collection of linear projections of a sparse signal contain enough information to reconstruct the signal. Compressed sensing makes it possible to develop new distributed coding algorithms for multisignal sets that use intra- and intersignals. This has given rise to distributed compressed sensing, which is a new concept that we call the joint sparsity of a signal set. A new model for sparse joint signals is introduced, which makes it possible to reconstruct multiple signals from incoherent projections through simultaneous tracking algorithms [DUA 05].

1

Shannon's Theorem in Classic Data Processing

1.1. Introduction

Lombardi *et al.* give a comprehensive explanation of Shannon information [LOM 16]. The question asked in their overview is: what is Shannon information? This chapter's approach seems useful in measuring or understanding compressed sensing and a prelude on Shannon's theorem is a necessary preliminary. Lombardi relies on numerous references: we highlight the work of Bell [BEL 57], Fisher [FIS 25], Khinchin [KHI 57], Kolmogorov [KOL 65, KOL 68, KOL 83] and MacKay [MAC 69] on data in general, without forgetting the depth of insight developed by Landauer [LAN 91, LAN 96].

It is traditionally accepted that Shannon's seminal work on the data point in math is the paper in which he introduces a specific formalism designed to solve some specific technical problems [SHA 48]. Shannon entropy is considered with a given system's statistical properties and the correlation between the states of two systems independently of their meaning or their semantic content. Shannon's theorem is the basic ingredient of digital communication studies. Currently, the philosophy of data has posed a number of problems linked to the concept of data. The possibility of unifying these data theorems, the question of the logic of information, the relationship between data and thermodynamics, the significance of quantum data and links between data and calculation are crucial to a complete understanding of data.

We could assume that all questions about the meaning and interpretation of Shannon data have had clear and undisputable answers. However, this is not the case.

1.2. Shannon's theorem

In his article on mathematical data theory [SHA 48], Shannon gave us specific results on the resources needed for optimal coding and error-free communication. Shannon's theorem was then formalized by Khinchin [KHI 57].

According to Shannon [SHA 48], a general communication system is formed of the following five parts:

– a source S that generates the message that should be received at the destination;

– a transmitter T that transmits the message generated at the source into a signal that should be transmitted. Where data are coded, the coding is also implanted into this system;

– a channel CH that is the medium used to transmit the signal from the transmitter to the receiver;

– a receiver R that reconstructs the message from the signal;

– a destination D that receives the message.

In Figure 1.1, we give an explicit schema of the system.

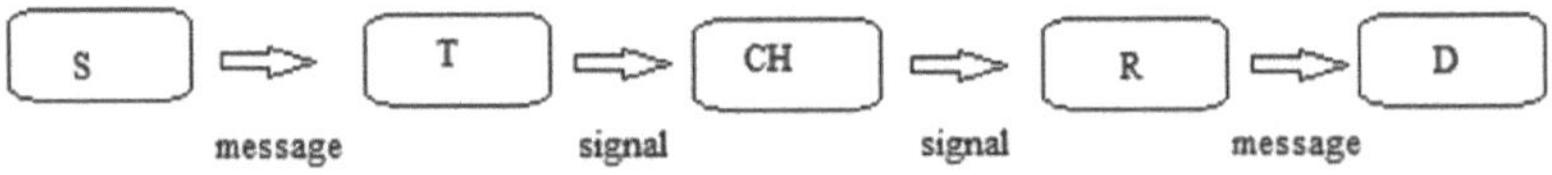

Figure 1.1. *Schema of the system*

The source S is a system with an interval of possible states $s_1,\dots s_n$ usually called letters, whose respective probabilities of occurrence are $p(s_1),\dots,p(s_n)$.

The quantity of information generated at the source through the occurrence of s_i can be defined as [SHA 48]:

$$I(s_i) = \log\left(\frac{1}{p(s_i)}\right) = -\log p(s_i) \qquad [1.1]$$

Since S produces sequences of states usually called messages, the source's entropy s is defined as [SHA 48]:

$$H(S) = \sum_{i=1}^{n} p(s_i)\log\left(\frac{1}{p(s_i)}\right) = -\sum_{i=1}^{n} p(s_i)\log p(s_i) \qquad [1.2]$$

By analogy, the destination D is a system with an interval of possible states $d_1,\dots, d_m$, with the respective probabilities $p(d_1), .. p(d_m)$. The quantity of data $I(d_j)$ received at the destination through the occurrence of d_j can be defined as [SHA 48]:

$$I(d_j) = \log\left(\frac{1}{p(d_j)}\right) = -\log p(d_j) \qquad [1.3]$$

And the entropy of the destination D is defined as [SHA 48]:

$$H(D) = \sum_{i=1}^{m} p(d_j)\log\left(\frac{1}{p(d_j)}\right) = -\sum_{i=1}^{m} p(d_j)\log p(d_j) \qquad [1.4]$$

In his article, Shannon explains that it is sensible to use the logarithm function to define entropies; it is useful on a practical level as most parameters in engineering vary linearly with the logarithm of the number of possibilities. This is intuitive, as measuring amplitudes by linear comparison with the units of measurement is more convenient as several operations, in terms of logarithms, are simpler in terms of the number of possibilities. If the base 2 is used, the resulting unit is called a bit, which is a contraction of "binary unit".

The relationship between the entropies of the source H(S) and the destination H(D) can be represented by Figure 1.2.

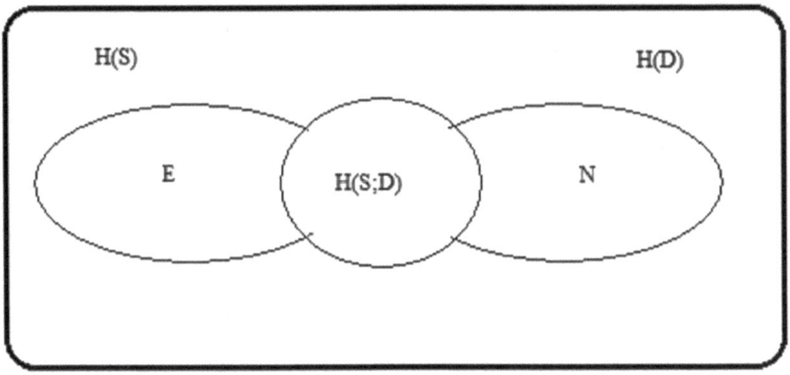

Figure 1.2. *Diagram of entropies*

The values of E and N are functions not only of the source and the destination, but also of the communication channel CH. Introducing the communication channel leads directly to the possibility of there being errors in the transmission process. The channel CH is defined by the matrix $[p(d_j/s_i)]$, where $p(d_j/s_i)$ is the conditional probability of d_j occurring in the destination D.

If s_i is occurrent in the source S, and the elements in any additional line up to 1, then on this basis E and N can be calculated as:

$$N = \sum_{i=1}^{n} p(s_i) \sum_{j=1}^{m} p(\frac{d_j}{s_i}) \log\left(\frac{1}{p\left(\frac{d_j}{s_i}\right)}\right) \quad [1.5]$$

$$E = \sum_{j=1}^{m} p(d_j) \sum_{i=1}^{n} p(\frac{s_i}{d_j}) \log\left(\frac{1}{p\left(\frac{s_i}{d_j}\right)}\right) \quad [1.6]$$

where $p(s_i/d_j)=p(d_j/s_i)p(s_i)/p(d_j)$. The capacity of the channel C is defined as:

$$C = \max_{p(s_i)} H(S;D)$$

where the maximum of all the possible distributions $p(s_i)$ at the source is taken. C measures the greatest quantity of information that can be transmitted on the channel CH.

1.3. Shannon's capacity channel

The two most important results obtained by Shannon are the theorems known as Shannon's first theorem and Shannon's second theorem. The first theorem, or Shannon's source coding theorem, has weak noise. This theorem stipulates that for long messages, the value of the source's entropy H(S) is equal to the average number of symbols needed to code a letter from the source using an ideal code. H(S) thus measures the optimal sensing of messages from the source.

The second theorem, or Shannon's noisy channel coding theorem, proves that the supposition is untrue so long as the rate of communication is kept lower than the channel's capacity. The channel's capacity is equal to the maximal rate at which information can be sent along the channel and can attain the destination with an extremely low probability of error.

Shannon's theorem is formulated for a noisy channel using work carried out by Litvin [LIT 09]. The proof of this theorem is based on the theory of optimal signal reception, in which the signal's intensity as a function of the noise is quite significant. Although the formulation contains the notions of the channel capacity and of message-source entropy as seen previously, they differ substantially from Shannon's formulation. The new formulation enables us to explain some instances where the data transmission conditions do not satisfy Shannon's theorem.

The channel's capacity is:

$$C = \max_{p(s)} \frac{1}{T_s} \iint p(s,x) \log \frac{p(s,x)}{p(s)p(x)} dsdx = \max_{p(s)} \frac{1}{T_s} [H(s) - H(s \mid x)] \quad [1.7]$$

Here, s and x are, respectively, the communication channel's input and output messages, p(x) and p(s) are the probability densities of the random variables, p(s,x) is the joint probability density of s and x, and

$H(s|x)$ is the unreliable channel. Equation [1.1] is maximized using the statistical messages. Optimal coding, the joint transmission of sequential messages and the increase in this transmission make it possible to decrease the probability of error.

1.4. Sampling

Signal processing methods have changed considerably over the course of recent decades. The number of operations that have shifted from analog to digital is constantly increasing. Sampling theory is the gate to the digital world; it is also the key that has unlocked this revolution.

Classic sampling theory or the Whittaker–Kotel'nikov–Shannon theorem on the reconstruction of signals from uniformly spaced samples assumes that the signal is limited by the band, i.e. with a spectrum in a bounded interval. This hypothesis was then reduced to a disjoint translation condition on the spectrum; Beaty and Dodson show that the disjoint translation condition is replaced by a nullity intersection condition, which is more natural on the translated spectrum. This condition reveals itself to be equivalent to Plancherel's isometric formula [PLA 36] when the spectrum has a finite measurement. Thus, extended multiband sampling theory has a logical structure similar to Fourier's classic analysis [BEA 04].

Madadi presents results on the Shannon information contained in the associated higher and lower values in a sequence of independent and identically distributed variables. He then establishes an interesting relationship between the Shannon information content of a random sample of fixed size and the Shannon information in the data consisting of maximal sequences [MAD 11].

Tamberg gives us a digital algorithm technique based on the Shannon sampling operators [TAM 10]. In the same vein, Boche studies the behavior of Shannon sampling series on Hardy spaces [BOC 09].

Mishali and co-authors introduce X sampling, called Xampling, which is a unified framework for acquiring and processing signals in a union of subspaces [MIS 11]. This framework has two main functions: analog compression, which reduces the input bandwidth before sampling with classic devices, followed by a nonlinear algorithm that detects the input subspace before conventional signal processing. A representative union mode with spectrally sparse signals serves as a test case for studying these Xampling functions. Three parameters have been taken for choosing the analog compression: robustness to model mismatch, the necessary material precision and software complexity. An exhaustive comparison between two sub-Nyquist acquisition strategies for a spectral broadband converter (MWC) was carried out in term of these measurements and it has been possible to draw operational conclusions about the choice of analog compression.

1.5. Shannon's theorem in cryptography

The most fundamental results in information theory are Shannon's theorems, which express the limits to a weak data compression and to data transmission on a noisy channel, as we saw previously. The proofs of both theorems are not obvious and they are rarely detailed in the literature. This lack of formal foundation is even more regrettable as the crucial results in data security rely solely on information theory; this is what is usually called unconditional security. Affeldt produced the first formal proofs of this source coding theory, which presents entropy as linked to lossless compression, and channel coding, which presents capacity as linked to weak communication on a noisy channel [AFF 14].

2

Shannon's Theorem in Quantum Data

2.1. Introduction

Another variant of Shannon's theorem, known as the Shannon–McMillan theorem, was the subject of work carried out by Ogata [OGA 13]. This chapter focuses mainly on the quantum aspects of data. The Shannon–McMillan theorem stipulates that an ergodic system has, as its solution, a set of typical series satisfying the asymptotic equipartition property. This theorem explains the meaning of entropy, which gives the dimension of the typical series. The asymptotic equipartition property can be seen as the equivalence of sets; it also makes it possible to give the potential compression rate in source coding theory.

Rastegin introduces a family of quantum measurements such as Shannon distinction. These measurements are defined using two categories and are linked to distinct parts of expression for mutual information; the changes in Ky Fan norms [FAN 55] and distribution distances resulting from the appearance of a partial trace are discussed. The upper and lower bounds of the quantities introduced are obtained in terms of the trace partitioning distances and Uhlmann's partial fidelity [UHL 00]. These inequalities provide a sort of generalization of the bounds well known from Shannon's distinction. The notion of cryptographic exponential indistinguishability for quantum states is reviewed. When a rapid exponential convergence is required, all the metrics induced by the invariant unitary norms are represented as being equivalent [RAS 11].

Discrepancies have appeared in Shannon's information theorem in quantum mechanics. A chapter on Shannon's theorem is required from a quantum perspective to understand some aspects of LiFi.

The fundamental research is owed mainly to Bennet and Shor [BEN 98] and to Deutsch and Hayden [DEU 00], who highlight the conceptual inadequacy of Shannon's information in measurements in quantum mechanics.

In an article published in *Nature*, Pusey *et al.* [PUS 12] argue that quantum states are the main mathematical objects in quantum theory and it is surprising that physicists have not been able to agree on what a quantum state really represents. One possibility is that a pure quantum state corresponds directly to reality; however, there is a suggestion that even a pure quantum state only represents knowledge or information on an aspect of reality. They rely on research on measurement carried out by Caves *et al.* [CAV 02]. Other research has been carried out by Preskill [PRE 98].

To summarize this research, we can say that quantum to classic mechanical equivalence is similar to the equivalence of complex numbers and real numbers. According to Deutsch and Hayden [DEU 00], causality in quantum mechanics is not information in Shannon's logic.

Wilde introduces Neumann entropy as the quantum generalization of Shannon entropy [WIL 16]. Quantum entropy gives meaning to the notion of qubit information. This notion is different to the physical notion of the qubit, which is a description of the quantum state of an electron or photon. Qubit information is a basic unit of quantum data measurement that determines how quantum information is present in a quantum system.

2.2. Definition of quantum entropy

The entropy of a system A is given as:

$$H(A)_\rho \overset{\text{def}}{=} -Tr\{\rho_A \log \rho_A\} \quad [2.1]$$

The density operator is defined as:

$$\rho_A \overset{\text{def}}{=} \sum_x p_X(x) \mid x >< x \mid_A \quad [2.2]$$

Quantum entropy $H(A)_\rho$ is the same as Shannon entropy: H(X) of a random variable X with a probability distribution $p_X(x)$.

2.3. The Holevo bound

The Holevo bound [HOL 73] is a particular case of the non-negativity of quantum discord in which ρ_{AB} is a quantum-classic state that we describe explicitly as:

$$\rho_{AB} = \sum_y p_Y(y) \rho_A^y \otimes |y >< y| Y \rho_{AB} = \sum_y p_Y(y) \rho_A^y \otimes |y >< y| Y \quad [2.3]$$

where each $\rho^y{}_A$ is a density operator, such that $\rho_A = \sum_y p_Y(y)\rho^y{}_A$. The Holevo bound stipulates that the mutual information of state ρ_{AB} in equation [2.3] is never smaller than the mutual information after the system is measured.

2.4. Quantum aspects of compressed sensing

These binary measurements are achieved through evaluating the inverse Gauss error function. Finally, these recuperations can be made on sparse vectors, in the sense that, with a high probability, a set of measurements and thresholds can be estimated with success for all the sparse vectors x in a Euclidean sphere of known radius [KNU 16].

Other work by Preskill [PRE 98] focuses on Shannon entropy and data sensing.

2.5. Compressibility and capacity

The compressibility of a quantum message can also be quantified, and, at least in this case where each letter in the message is a pure quantum state, the answer is directly analogous to that found by Shannon.

A quantum channel is one that transmits qubits rather than classic bits. The capacity of such a channel is revealed to be a subtle concept and a variety of important questions remain open. Several different types of channel capacity can be formulated and studied; in particular, it is important to distinguish between the quantity of classic information and the quantity of quantum information that can be transmitted reliably on a quantum channel.

One of Shannon's great realizations was that a random code could reach the maximum communication speed possible on a classic channel.

For some decades, researchers have constructed data models for molecular interaction to predict the new molecules' properties. These models take the form of a potential function, equations that can be used to predict the interaction's molecular energy. Recently, there has been a revolution in statistical approaches that could help to solve these problems and aid an approach hinging on data at new levels. At the center of this approach, compressed sensing is a technique for finding sparse solutions on unspecified linear systems, i.e. solutions that do not have these surfeits. This is done using a regularization schema, which systematically avoids surfeiting by penalizing overly complex models [PAN 11].

Compressed sensing has been used lastly to solve certain intrication problems. As evidence, we give Plenio's work [PLE 05]. The exact representation of an intricate state of N parts requires the specification of a number of states that increases exponentially in N. Most of the numbers are small, in fact exponentially small and are not relevant for calculating most physical sizes. Consequently, effective methods for determining physical sizes should take account of this fact and measure only what is really necessary. Solving these problems will bring about a future solution for ion trapping.

Howland used a dual-pixel detection camera for compressed sensing to characterize effectively, at high resolution, the spatially intricated fields that are produced through a low parametric spontaneous conversion. This technique uses sparseness in spatial correlation between the intricated photons to improve the sensing time on the raster scan by a scale factor of up to $n^2/\log n$ for n dimensional images. Howland visualizes resolutions of up to 1,024 dimensions per detector and he demonstrates a channel capacity of 8.4 bits per photon. By comparing the classic mutual information of the intricated photons in conjugate bases, he demonstrates violation of the entropic separability criteria EPR (Einstein-Podolsky-Rosen) for all resolutions measured. More generally, his result indicates that detection by compression can be particularly effective for higher order measurements on correlated systems [HOW 13].

If a measurement provides information on the quantum system's position statistics, a later measurement can provide information on the system's momentum statistics. Understanding of information is incorporated into the entropic formulation of the uncertainty principle.

Traditionally, the uncertainty relationships correspond to the resolution limits; increasing a detector's positioning sensitivity decreases its momentum sensitivity and vice versa. However, this is not generally necessary; for example, position data can be extracted at the cost of noise in the momentum. Using random and partial position projections followed by a high momentum measurement, we effectively determine the distributions of the transverse position and the transverse speed of an unknown optical field with a single set of measurements. The momentum distribution is imaged directly, whereas the position distribution is recovered using detection by compression. We do not violate the uncertainty relationships at any momentum; rather, we economize on usage of the data that we obtain [HOW 16].

Precise establishment of the state of large-scale quantum systems is an important tool in quantum information science, but the large number of unknown parameters hinders rapid characterization of such states and reconstruction procedures can become excessively time consuming. Compressed sensing is a procedure for solving inverse problems by incorporating prior knowledge on the form of the solution; it is an interesting alternative to the problem of characterizing the high-dimensional quantum state. Using a modified version of compression detection that incorporates the principles of singular-value threshold, Tonolini and co-workers reconstitute the density matrix of a high-dimension intricate two-photon system. The algorithm that they develop is fast, takes little calculation and is applicable to a broad range of quantum states. They thus demonstrate that compressed sensing is an effective technique for measuring the state of large-scale quantum systems [TON 14].

Lum and co-workers demonstrate how to effectively obtain high-dimension imagery through the compressed sensing of a two-photon probability distribution. Their method uses the fast Hadamard transform at the root of compressed sensing to acquire the spatial distribution. They give, in detail, the operations needed to make it possible for the fast transform based on the matrix-vector to reconstruct an image with 16.8 million dimensions in less than 10 min. In a subspace of this image, there is a biphoton probability distribution. Moreover, they show how marginal distributions can aid precision in reconstructing spatial distribution [LUM 15].

Well-controlled quantum devices with their increasing system size face a new obstacle hindering the development of quantum technologies: the stress of quantum tomography, the characterization of processes and states within a quantum device, breaks down unfavorably to the point that the state of the point systems can no longer be processed. Quantum compressed sensing reduces this problem by reconstructing the state from an incomplete set of observables. Riofrio shows an experimental implementation of the compressed tomography of a system with seven qubits, which constitutes the greatest discovery to date and he introduces new digital methods in order to extend reconstruction to this dimension. Originally, compressed sensing was recommended for density matrices with few non-null eigenvalues. Riofrio argues that low-ranking estimates provided by compressed sensing can be appropriated even in general cases. The reason for this is that statistical noise often makes it possible only to reconstitute the main eigenvectors reliably: he shows that the remaining eigenvectors behave in a way compatible with a random matrix model that carries no information on the real state. He reports reconstruction of quantum states from a topological color code with seven qubits, prepared in a trapped-ion architecture, based on topographically incomplete data involving 127 Pauli base measurement parameters, each repeated 100 times [RIO 17].

2.6. Conclusion

Experimental realization and characterization of entanglement become extremely difficult for systems with many particles. The main difficulty lies in manipulating and detecting the quantum state of individual particles as well as in controlling interaction between them. Compressed sensing with reconstruction methods can bring about a noticeable improvement in data knowledge even in cases where particles are lost.

3

Sparse Signals and Compressed Sensing

3.1. Introduction

Compressed sensing arose from the need to process very complex problems in signal and image processing. The idea is to use the image's original spatial redundancy to extract an image summary containing the patches of texture most representative of the image, then to use this compact representation to reconstruct the initial image. The notion of sparsity has emerged as a fundamental requirement in recent years. A signal is said to be sparse on a basis if it can be described by a low number of non-null coefficients on this basis.

3.2. Definition of sparse decomposition

Consider a system of equations covering a matrix D of indeterminate dimensions (n, n), as well as a signal written α on which we apply a constraint known as sparse; this can be represented by: $x = D\alpha$, $x \in \mathbb{R}^m$, and the signal $\alpha \in \mathbb{R}^n$, which is generally called D, dictionary.

Sparsity assumes that the number of non-null components α is low and the majority of the components are null or insignificant. Unlike another technique called principal components decomposition

[KAL 02, KOU 02, AI 15], the base vectors are not considered to be orthogonal.

The problem of sparse decomposition as reported in the literature is reduced to the following equation:

$$\min_{\alpha\in\mathbb{R}^n}\|\alpha\|_0 \text{ such that } x = D\alpha \qquad [3.1]$$

where $\|\alpha\|0$ is a pseudo norm l_0 that includes the number of non-null components of $\alpha = [\alpha_1, \ldots, \alpha_p]^T$. A convex relaxation can be obtained by taking norm l_1 instead of norm l_0, where:

$$\|\alpha\|_1 = \sum_{i=1}^{p}|\alpha_i| \qquad [3.2]$$

Norm l_1 has no preference between the large and small components. However, norm l_2 much prefers the small components, whose square becomes negligible, to the large components, whose square becomes enormous. Norm l_1 makes it possible to have some large components if this is required, whereas norm l_2 forbids it.

3.3. Sparse signals

Defining the signal model, we consider a signal with actual values without loss in general, and the spatial dependencies are also considered. We replace the variable r with another convenient variable: time, frequency and the wave vector.

The signal $f(r)$ can be represented by a development of basis functions $\gamma n(r)$ as follows:

$$f(r) = \sum_{n=1}^{N} f_n \gamma_{n(r)} \qquad [3.3]$$

If f is a multidimensional vector with the dimension N such that:

$$f = \{f_n \in \mathbb{R}, n = 1, \ldots, N\}$$

the image y(r) is sent back to f(r) through:

$$y(r) = £\{f(r)\} \qquad [3.4]$$

£{being the linear observation operator. We assume that the signal y(r) is acquired though the following measurement operation:

$$y_m = \langle y(r), \varphi_m(r) \rangle, \text{ where } m = 1,\ldots.., M \quad [3.5]$$

$\langle . \rangle$ is the scalar product, y(r) and $\varphi_m(r)$ are acquisitions with the wave form ψ, (for example Dirac delta functions) with $M \leq N$.

3.4. Compressed sensing and optimization

Recht shows that if a certain restricted isometry property holds for the linear transformation defining the constraints, the rank minimum solution can be recovered by solving a convex optimization, i.e. minimizing the nuclear norm on the given affine space. He presents several random sets or equations where the restricted isometry property holds with overwhelming probability on the condition that the codimension of the subspace is large enough. The techniques used in his analysis are similar to those of compressed sensing. He discusses how affine rank minimization generalizes the preexisting concept and introduces a dictionary revealing concepts of cardinality minimization and rank minimization. He also discusses several algorithmic approaches to minimizing the nuclear norm and illustrates his results with digital examples [REC 10].

The context of compressed sensing makes it possible to recover a sparse signal from a small number of projections on random vectors. Andrecut suggests a new method of extending compressed sensing, applied to sparse signals in space time. This procedure consists, in a competitive set, of compressed sensing devices. He shows that this approach leads to a substantial improvement in performance. Moreover, he suggests a simplified version of compressed sensing, in which random projections are replaced by a cyclic correlation with a single random vector. He shows that the performances can be improved still further using a simulated annealing algorithm [AND 07].

A symmetrical version of the Neyman–Pearson test has been developed to differentiate between series of hypotheses and to cover a new formulation of the problem of estimating parameters according to sets of finite data. Such problems can occur in the distributed acquisition and localization of problems in networks of sensors, where data from the sensors should be compressed to take account of the constraints in communication. In this context, it is natural to focus on methods that balance coarse resolution of estimations in order to reach greater reliability [BOR 04].

3.5. Reconstructing the sparse signal

Compressed sensing is a signal processing method that acquires data directly in a compressed form. This makes it possible for fewer measurements to be considered necessary to record a signal, which permits faster or more precise measurement protocols in a broad range of applications. Krzakala has recently suggested a strategy that makes it possible to obtain compressed sensing at acquisition rates approaching the optimal theoretical limits. He gives a deeper presentation of his approach and a second introduction to many new results. He introduces the probabilistic approach to reconstruction and he discusses its optimality and robustness. He details how to determine messages passing algorithms for reconstructing and maximizing attempts at learning the signal model's parameters. Afterward, he develops asymptotic analysis of corresponding phase diagrams with and without measurement noise for different signal distributions and discusses the best possible reconstruction performances [KRZ 12].

Takeda and Kabashima provide a system for exploring the limits for reconstructing compressed sensing by minimizing the general cost function in the context of random measurement constraints for the sources of correlated generic signals. The system is based on the statistical mechanical replica method for processing random systems. A simple but not insignificant example is as follows: we apply the system to a sparse autoregressive model, where the main differences in the input signals of the correlated time series are sparse, and we evaluate the critical compression ratio for a perfect reconstruction. The results agree well with a digital experiment for signal reconstruction [TAK 11, TAK 13].

The context of 1-bit compressed sensing makes it possible to recover a sparse x vector from the information from each input of its linear sign transformation. Removing the amplitude information makes it possible to reduce the quantity of data considerably, which is very beneficial in practical applications. In this chapter, we introduce a Bayesian approach for signal reconstruction for 1-bit compressed detection and analyzing its typical performance using statistical mechanics. Through basic configuration, we consider the example of the measurement matrix ϕ that has i.i.d (independent and identically distributed) inputs and measurement ψ. The measurement matrix and measurement have weak noise. By using the replica method, we show that the Bayesian approach permits better reconstruction of the minimization norm l_1. We also test a message passing the signal reconstruction algorithm on the basis of belief propagation. The results of the digital experiments conform to those of theoretical analysis [XU 14].

Compressed sensing is a data compression technique that uses new convex optimization methods [CAN 09]. Compressed sensing uses the correlation inherent in a particular input data set x to compress these data by means of quasi-random matrices [TSA 06]. If the compression matrix and the original x data have certain properties, x can be reconstructed from its compressed version Y with a high probability by minimizing a distance metric on a solution space. Compressed sensing was initially developed for the efficient storage and compression of digital images, which show a high spatial correlation [LUS 07]. Recently, there has been growing interest in these techniques from telecommunications signal processing companies [DUA 11]. Unlike classic approaches, where the data are first compressed then transmitted to a given destination, with compressed sensing the compression phase can be carried out in conjunction with the data transmission. Compressing data before transmission to the data collection point (DCP) requires first knowing the correlation properties of the input signal on the entire network [XIA 11] (or a large part of it) and this involves high transmission costs. With compressed sensing, the content of the packets can be mixed as they are rerouted to the DCP [HAU 08]. Under some conditions, compressed sensing enables the reconstruction of all the sensor readers in networks by using far fewer transmissions than routing or aggregation systems. These characteristics make compressed sensing a very promising asset for the joint acquisition and aggregation of data from devices distributed in a network of sensors without multihop wireless [LIU 15].

In this section, we also recall the main concepts of the Nesterov algorithm for effectively solving the convex optimization problems in compressed sensing [NES 94]. This point of view is discussed by Bach *et al.* [BAC 12].

4

Compressed Sensing and the Fourier Transform

4.1. Introduction

The fast Fourier transform (FFT) has been applied to sparse signals. The aim is to calculate only the significant coefficients in the linear time frequency domain.

Hassanieh introduces an application consisting of a receiver that can acquire a passband six times greater than detection in real time. Its sampling rate makes it possible to use components generally used in WiFi receivers [HAS 14].

Magnetic resonance spectroscopy (MRS) detects the biochemical content of each voxel in the brain and can be used to discover the biological markers of disease. It has been shown that data processing using MRS combined with the sparse Fourier transform (SFT) improves the quality of the image and reduces the patient's waiting time by three times [DO 12].

In photography, the light field can refocus and/or modify its point of view by postprocessing. Reconstructing images from light fields using the sparse FFT reduces sampling demands and proves the quality of the reconstruction [AND 07].

The sparse FFT has found an interesting application in GPS; it makes it possible to reduce the time and power needed to lock the receiver into its place [HAS 12].

With the new algorithm, called the SFT, the data flux can be treated 10–100 times faster than possible with FFT. The acceleration can be produced because the information in which we are most interested has a great deal of structure: this information has no random noise. These meaningful signals typically have only a fraction of the possible values that a signal can take; the technical term used is to say that the information is "sparse". Because the SFT algorithm does not aim to work with all the possible data flows, it can undergo some shortcuts that are not available elsewhere. In theory, an algorithm that can only process sparse signals is much more limited than the FFT.

We consider the Fourier transform in this sparse instance: given a complex vector x of length n, and a parameter k, estimate the largest k (in amplitude) as a coefficient of the Fourier transform of x. This problem is of major interest in several domains: signal processing, audio/image/video compression and the learning theory. A new algorithm for this problem is suggested [RAO 11]. The algorithm takes some of the techniques for processing digital signals, particularly Gaussian filters and Dolph-Chebyshev. Unlike a typical approach to this problem, the algorithm is not iterative. In other words, instead of estimating "large" coefficients, it subtracts them recursively, and identifies and estimates the largest coefficients k. It is a diffusion algorithm. The resulting algorithm is structurally simpler than its predecessors. In consequence, we are able to extend considerably the range of sparsity, k, for which the algorithm is faster than the FFT, both in theory and practice [KUM 94].

4.2. Fast SFT

The discrete Fourier transform (DFT) is one of the most important and most well-used calculation tasks. It applications are broad and include signal processing, communications and/or audio/video image compression. Consequently, fast algorithms for DFT are invaluable. At the time of writing, the fastest algorithm is the FFT, which calculates the DFT transform of an n-dimensional signal in O (n log n). The existence of DFT algorithms faster than the FFT is one of the central questions in algorithm theory [KUM 94].

A general algorithm for calculating the DFT exactly should take a length of time at least proportional to its output size n. In many applications, however, most of a signal's Fourier coefficients are less than or equal to zero, i.e. the output of the DFT is sparse. This is the case for video signals, where a typical block of 8 × 8 in a video frame has on average seven non-negligible frequency coefficients (i.e. 89% of coefficients are negligible). For sparse signals, the lower limit Ω (n) of the complexity of the DFT no longer holds. If a signal has a small number k of non-null Fourier coefficients, the Fourier transform can be represented succinctly using only coefficients of k. Consequently, for these signals, we can find DFT algorithms whose execution is sublinear in the size of the signal (n).

Here, we show several new results for the SFT:

– an O (k log n) – algorithm time for the sparse case, exactly k;

– an O (k log n log (n/k)) – algorithm time for the general case;

– an Ω (k log (n/k)/loglog n) lower limit of the sample complexity.

Both algorithms improve with the FFT, for any k = o (n). Moreover, if we assume that the FFT is optimal, the algorithm for the sparse instance, exactly k, is optimal. According to the same hypothesis, the result for the general case is, in the longest instance, an n-factor loglog, which is far from an optimal execution for the case with "large" sparsity k = n/log n.

Moreover, we show that the SFT has an impact on many practical applications in the domains of WiFi communication, medical imaging and computing photography.

Candés and Wakin [CAN 08b] note that there are superficial similarities between compressed sensing and coding theory, more specifically in the theory and practice of Reed Solomon codes. In fact, it is well known that it is possible to adapt ideas from coding theory to establish what follows.

We can reconstruct, uniquely, any sparse signal S from its first Fourier coefficients, $y_k = \sum_{t=0}^{n-1} x_t\, e^{-i2\pi\frac{kt}{n}}$, k=0,1,2,.., 2S-1, or from any set of 2S consecutive frequencies (the calculation cost is essentially that of solving a system S × S from a Toeplitz system with an FFT with n points). Does this mean that we can use this technique to detect compressed signals? The answer is no. This is mainly due to the following two reasons.

First, Reed Solomon decoding is an algebraic technique that cannot process non-sparse signals; decoding is supported by a polynomial formulation.

Second, the problem of finding support for an even signal when the signal is exactly sparse from its prime Fourier coefficients is an ill-posed problem. This is exactly the same as the problem of extrapolating a higher degree polynomial from a small number of strongly grouped values. Small perturbations in these coefficients will give completely different responses with finite precision data and a reliable estimation of the support is practically impossible, whereas purely algebraic methods ignore the conditioning of information operators with well-conditioned matrices that are essential for precise estimation. This is a central problem in compressed sensing in which the restricted isometry property plays an obvious role.

In compressed sensing, structured matrices following a Toeplitz system have been discussed by Bajwa *et al.* [BAJ 07]. We emphasize that if A is a random matrix with inputs drawn independently from some probability distributions, the exact recovery of x from these observations can be guaranteed with high probability. The use of Toeplitz matrices in compressed sensing presents several advantages: generation of only 0(n) independent random variables and multiplication with Toeplitz matrices can be carried out effectively using the FFT, allowing reconstruction and faster acquisition by algorithms; the matrices structured by Toeplitz appear naturally in some application domains, such as system identification.

Hassanieh *et al.* [HAS 12] propose a simple and practical algorithm for the SFT. The algorithm uses digital signal processing techniques, especially Gaussian and Dolph-Chebyshev filters. Unlike the classic approach, the algorithm is not iterative, i.e. instead of estimating large coefficients by subtracting them, he identifies and estimates the k largest coefficients all together, in a way analogous to streaming algorithms. The resulting algorithm is structurally simpler than its predecessors. Consequently, it is able to considerably extend the range of sparsity, k, for which the algorithm is faster than the FFT as much as theory as in practice.

Gilbert *et al.* give an almost sparse representation of the Fourier optimal through sampling [GIL 02]. This algorithm makes it possible to find a Fourier representation R in B terms for a discreet signal A of length N, such that $||A - R||_2^2$ is within $(1+\varepsilon)$ for the better $||A - R_{opt}||_2^2$. The algorithm makes it possible to permit access A by listing its values on a set of samples T [0, N] chosen at random from among a distribution of our choice, i.e. we sample non-adaptatively. The total cost of the algorithm is polynomial in B log (N) log (M) /e, where M is the ratio between the greatest and smallest digital quantity encountered, which involves a similar limit for the number of samples.

To conclude this section on the SFT, we will not fail to mention the work carried out by Mallat on signal processing by wavelets in the context of sparsity [MAL 09].

4.3. Random matrices in compressed sensing

Covariance test structures are of great benefit in many domains of statistical analysis and the construction of compressed detection matrices is a significant problem in signal processing. Motivated by these applications, Cai, in an article, studies the laws limiting the coherence of a random matrix nxp in the dimensional framework where p can be much larger than n, so much so that both the law of large numbers and distribution limit are determined. We then study how to test the covariance matrix of a dimensional Gaussian distribution that includes independence tests as a particular case. The limiting laws of the coherence of the data matrix play a critical role in building the test. We also apply asymptotic results to building compressed detection matrices [CAI 11].

Literature on processing sparse signals often uses random detection matrices to obtain guarantees of performance. Unfortunately, in the real world, detection matrices do not always result from random processes. It is therefore desirable to evaluate whether an arbitrary matrix, or matrix framework, is adapted for sparse signals. To this end, Mixon examines two parameters that measure the coherence of a framework: the worst case and the average consistency. He first provides several examples of frames that have a small spectral norm, worst case, consistency and mean consistency. Then it presents a new lower bound of the worst case consistency and compares it to the bound Welch. Subsequently, he suggested an algorithm that reduces the average coherence of a frame without modifying its spectral norm or worst case, the coherence. Finally, he uses the worst case and the average frame; in contrast with the restricted isometry property, this makes it possible to recover more optimal and probabilistic guarantees regarding both the detection of sparse signals and reconstruction in the presence of noise. This contrasts with recent results which do not guarantee recovery of the weak signal noise from arbitrary images, and which moreover assume independence through the signal's non-zero inputs in one sense, which requires a weak average coherence that replaces the need for such a hypothesis [MIX 11].

Candés and Tao introduce the notion of the uniform uncertainty principle. This principle establishes that an acquisition matrix KxK, written as Φ, obeys the restricted isometry hypothesis. If Φ_T, TC{1,….,N}, and if Kx$|T|$ the submatrix obtained by extracting the columns of Φ corresponding to indices in T, then the restricted isometry constant in s written as δ_s for Φ is the smallest quantity defined by [CAN 05]:

$$(1 - \delta_S)\|c\|_{l2}^2 \leq \|\phi_T c\|_{l2}^2 \leq (1 + \delta_S)\|c\|_{l2}^2 \qquad [4.1]$$

For all the subsets T with $|T| \leq S$ of the coefficients of sequences $(c_j)_{j\in T}$. This property essentially demands that each set of columns of the acquisition matrix Φ are approximately orthogonal, and so the phenomenon of exact recovery occurs.

We can assume that all of this would be interesting if we designed a matrix permitting us to recover as many inputs of x as possible with as few K measurements. According to the theory given by Candés and Tao [CAN 05], we would like the condition $\delta_2 s + \theta_{s,2s} < 1$ to be obtained for large values of s, ideally in the order of K. This poses a problem with design. How to design a matrix Φ that is a collection of N vectors in K dimensions – such that any subset of columns of a size in the region of s is orthogonal? And for what values of s is this possible?

While it can be difficult to expose a matrix that obeys the uniform uncertainty principle for very large values of s, we know that trivial random constructs will do this with high probability.

We discuss four cases: Gaussian measurements, binary measurements, Fourier measurements and incoherent measurements.

4.3.1. *Gaussian measurements*

Here, we assume that the inputs of K by N of the acquisition matrix Φ are sampled independently from normal distribution with a null average and a variance equal to 1/K.

Therefore, if:

$$s \leq C \cdot K / \log (N/K), \qquad [4.2]$$

s obeys the conditions of the theorems given by Candés, Romberg and Tao [CAN 06a] with a probability $1 - O(e^{-\gamma N})$ for $\gamma > 0$. The evidence uses known concentration results on singular values of Gaussian matrices [SZA 91].

4.3.2. *Binary measurements*

Assume that the inputs for the acquisition matrix K by N of the acquisition matrix Φ are sampled independently from symmetrical

Bernoulli distribution P ($\Phi_{ki} = \pm 1\sqrt{K}$) = 1/2. We thus conjecture that the conditions of the theorems given by Candés [CAN 06] are satisfied with a probability 1 – O ($e^{-\gamma N}$) for $\gamma > 0$.

4.3.3. *Fourier measurements*

Now assume that Φ is a partial Fourier matrix obtained by randomly selecting K lines uniformly as above, and by renormalizing the columns following a unitary norm. Candès and Tao [CAN 06b] have shown that the previous conditions are obtained with a high probability if $s \leq C{\cdot}K/(\log N)^6$.

Recently, Rudelson and Vershynin [RUD 06] have improved this result and have obtained $S \leq C{\cdot}K/(\log N)^4$. This result is non-trivial and uses sophisticated techniques of functional analysis and probability in Banach spaces. It is conjectured that $S \leq C{\cdot}K/\log N$ remains valid.

4.3.4. *Incoherent measurements*

Now assume that Φ is obtained by selecting K lines uniformly and randomly from an orthonormal matrix U N by N and renormalizing the columns so that they are standardized units. As before, we could consider a matrix U as the matrix $\Phi\Psi^*$ that maps the object of the domain Ψ onto domain Φ. So, according to [CAN 06b] and [RUD 06], the uniform uncertainty principle is obtained for incomplete Fourier matrices:

In this case:

$$s \leq C.\frac{1}{\mu^2}\frac{K}{(logN)^4} \qquad [4.3]$$

where:

$$\mu = \sqrt{N}\, max_{i,j}\left|U_{i,j}\right| \qquad [4.4]$$

Note that for the Fourier matrix, $\mu = 1$ which gives the result in a special case.

If $U = \Phi\Psi^*$, then μ will be written as:

$$\mu = \sqrt{N} \max_{i,j} |\langle \varphi\_(i,), \psi_j \rangle| \quad [4.5]$$

This expression represents the mutual coherence between the measurement base Φ and the sparsity base Ψ. The greater the incoherence of the measurement/sparsity pairing, the lower the number of measurements needed.

4.4. Conclusion

The random matrices that we have introduced and which are used in compressed sensing should meet the restricted isometry property; among these, we distinguish random or Gaussian matrices. The choice of random matrix depends on the algorithmic complexity that results from them. We can find a panel of random matrices verifying the restricted isometry property with high probability, and weak coherence with the sparsity matrix. Some of these matrices are built from Fourier matrices, which have the advantage of being diagonal.

5

Compressed Sensing and Entanglement

5.1. Introduction

Intuitively, if a density operator has a small rank, it should be easier to estimate from experimental data, since in this case only a few eigenvectors must be taken into consideration. Flammia shows two complementary results that confirm this intuition. First of all, he shows that a low-rank density matrix can be estimated using fewer copies of the state, i.e. the tomography's sample complexity decreases with the rank. Second, he shows that the unknown low-order states can be reconstructed from an incomplete set of measurements using compressed sensing techniques and matrix completion. These techniques use simple Pauli measurements, and their production can be certified without making any hypothesis on the unknown state. Then, Flammia presents a new theoretical analysis of compressed tomography based on the restricted isometry property for low-rank matrices. Because of these tools, he obtained nearly optimal error bounds for the realistic situation where the data contain noise due to finite statistic, and the density matrix is of full rank with decomposing eigenvalues. He generally obtains upper limits on tomography's compressed sample complexity, which also correspond with its lower bounds on the sample complexity of any procedure using Pauli measurement adaptation sequences [FLA 12].

5.2. Uncertainty relationships

In his article on uncertainty relationships for shift-invariant analog signals, Eldar [ELD 09] adapts the notion of coherence defined for finite dictionaries to infinite representations of the shift-invariant. This class of signals is rich enough to include several special, interesting cases such as multiband signals and splines. She shows how to find a sparse decomposition on an overcomplete dictionary by solving a convex optimization problem.

According to Eldar [ELD 10], the uncertainty relationships postulated by the research of Weyl [WEY 50] and Heisenberg [HEI 49], which shows that a signal can simultaneously be localized in time and frequency, have been extended by the research of Landau *et al.* [LAN 72] and lately by Donoho and Stark [DON 89] in the case of signals that are not limited to concentration on a single interval. In an harmonic analysis, he imposes limits on the time frequency resolution. The main benefit of representations in redundant dictionaries comes from the fact that the flexibility these systems offer can lead to decompositions that are extremely sparse, using only some elements of the dictionary.

However, finding an infrequent expansion in practice is a general and difficult combinatorial optimization problem. Two basic questions at the heart of full representations are as follows: what is the smallest number of elements in the dictionary needed to represent a given signal and how can we find the IT development effectively?

5.3. Phantom images

A phenomenon that has intrigued physicists for some time is the curious creation of "phantom images"; it was thought that this resulted from intrication. This phenomenon can be explained by compressed sensing. Bromberg gives an experimental explanation of this phenomenon [BRO 09].

The precise establishment of large-scale quantum systems is an important tool in quantum data science; however, the large number of unknown parameters hinders fast characterisation of such states, and the reconstruction procedures can become prohibitive in time. Compressed sensing is a procedure for solving inverse problems by incorporating existing knowledge on the form of the solution; it offers an interesting alternative to the quantum state problem of large dimensional characterization. By using a modified version of compressed sensing that incorporates the principles of singular value thresholding, we reconstruct the large-dimension density matrix from a biphoton intricated system. The dimension of each photon is d, equal to 17, which corresponds to a system of 83.521 unknown parameters. Precise reconstruction is obtained with around 2,500 measurements, i.e. 3% of the total number of unknown parameters in the state. The algorithm that we develop is fast, costs little in information and is applicable to a wide range of quantum states, thus demonstrating that compressed sensing is an effective technique for measuring the states of large-scale quantum systems [TON 14].

5.4. Sparsity in practice

We implement a double-pixel compression-detection camera to characterize effectively, at high resolution, the spatially intricated fields that are produced through spontaneous parametric conversion. This technique is based on sparsity in the spatial correlations between the intricated photons to improve the acquisition time of scanning by a sale factor up to $n^2/\log(n)$ for images of n dimension. Imaging occurs at resolutions of up to 1,024 dimensions per detector and demonstrates a channel capacity of 8.4 bits per photon. By comparing the classic mutual information of intricated photons in conjugated bases, we violate the Einstein–Podolsky–Rosen criterion, which is a separability criterion for all the resolutions measured. More broadly, our result indicates that compressed sensing can be particularly effective for higher order measurements on correlated systems [HOW 13].

To illustrate this chapter with a practical application, we outline quantum encryption software.

5.5. Chaotic and quantum encryption software

5.5.1. *Introduction*

Many systems have been developed on our interface, which is created in Delphi technique. In order to make this interface more supple and usable, we separate it into several programs. The first will act as a main menu interface and it is through this that the user will interact with the software; the second illustrates some systems in classic cryptography, while the third and fourth parts are reserved for chaotic and quantum cryptography, respectively.

5.5.2. *The software's content*

This software contains several cryptographic techniques, memory results and tasks aiding use of this software such as help and installation file.

In the part reserved for classic cryptography, it contains the following algorithms.

Des, 3Des, Blowfish, Twofish, RC2, RC4, CAST, cost, Rijndael

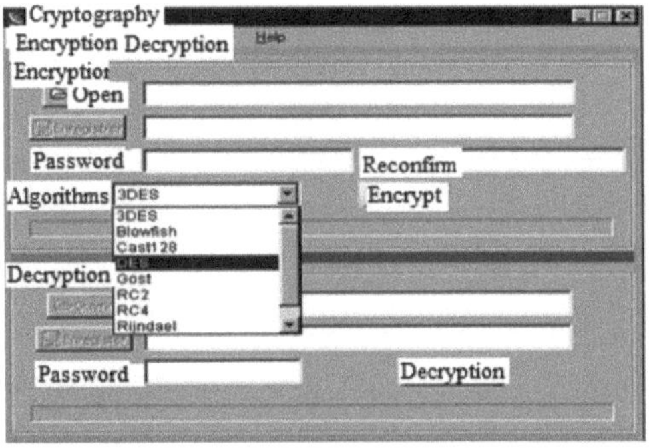

Figure 5.1. *Classic cryptography menu (summary)*

The part reserved for chaotic cryptography contains:

– results from this section (sensitivity to initial conditions, bifurcation diagram, Feigenbaum diagram);

– two methods for chaotic cryptography and their analysis;

– CKBA and their analysis;

– BRIE and their analysis;

– improvement of CKBA and BRIE using data compression:

- improvement of classic techniques using chaotic orbits.

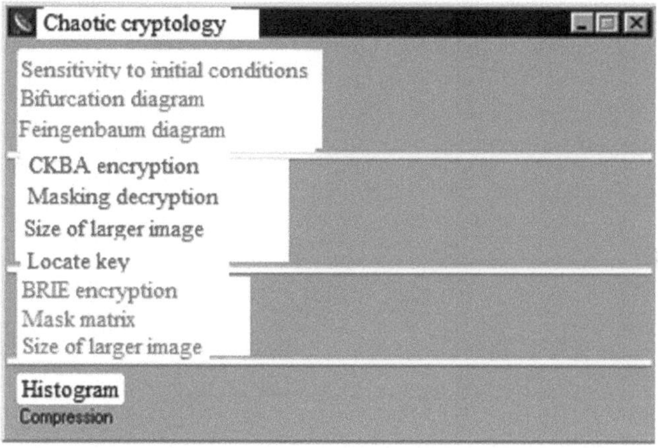

Figure 5.2. *Chaotic cryptography menu (summary)*

Figure 5.3. *CKBA*

Finally, we find the part reserved for quantum cryptography for this part's results.

– value of the Wigner function depending on losses;

– mutual information depending on equivalent losses;

– evolution of mutual information depending on equivalent losses;

– mutual information depending on isolated losses.

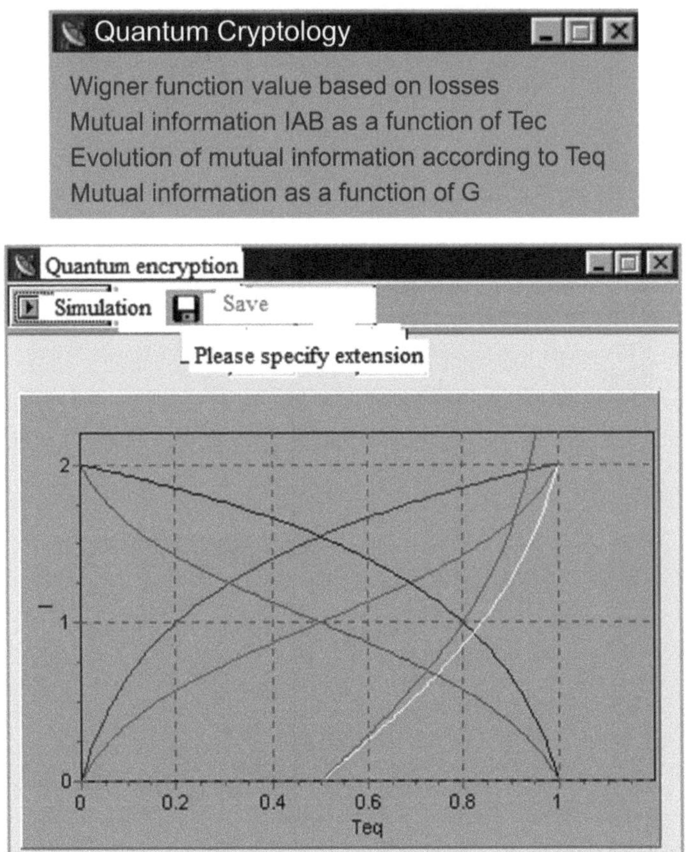

Figure 5.4. *Quantum cryptography menu (summary)*

6

[WITHDRAWN] Compressed Sensing and Intelligent LiFi Systems

This article has been withdrawn: please see Elsevier Policy on Article Withdrawal (http://www.elsevier.com/locate/withdrawalpolicy). This chapter has been removed due to plagiarism issue regarding some of the material, and has been removed with full knowledge of the authors.

7

Compressed Sensing in LiFi Systems in Mobile Communications and Cryptography

7.1. Introduction

LiFi technology is dependent on the type of diode or lamp used. Therefore, Yang suggests a PIN photodiode. The LiFi–EC–PD is manufactured on a bandwidth at –3 dB of 20 GHz and a response of around 0.4 A/W [YAN 04].

Gilliard has developed a LiFi lamp with high efficiency and a large white light, with a life span of more than 30,000 h. This lamp is also characterized by having 150 lumens/W. This lamp can also be controlled digitally [GIL 10, GIL 11].

Rufo *et al.* have implanted a prototype of a wireless optical system based on LED lamps (electroluminescent lamp diode) with a visible white light. This prototype makes it possible to diffuse a video at a rate of 2 Mbp/s. This technology, called visible light communication, presents several advantages such as electromagnetic interface robustness, visual security and network security. This technology is compatible with radiofrequency networks, WiFi, Bluetooth, etc. [RUF 10]. The receiver used is configured in infrared with 115 kb/s. Another receiver is designed for the same purpose. It is structured around IEEE 801.11, fixing the SISO IEEE 801.11 by Bin [BIN 11].

Li designed and created a type of LCD based on a microprocessor chip ARM9 [LI 11]. A diagram of the system is shown in Figure 7.1.

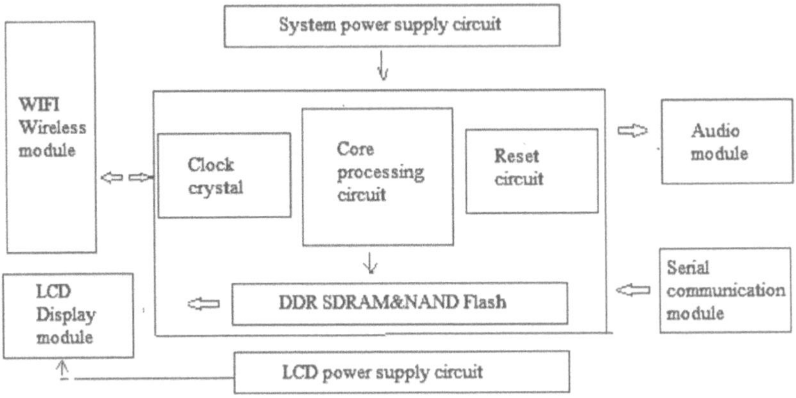

Figure 7.1. *Diagram of the system*

This technology hinging on visible light communication is also tackled by Pujapanda in his article on intelligent lighting, a technique that has attracted attention in particular due to its ecological qualities. Successful trials have been carried out by Pujapanda, which show that integrating visible light communication with high-power communications forms the start of access to broadband for domestic networks; they provide efficiency and low lighting costs [PUJ 13]. Along the same lines, Sung has designed a light control system with LEDs; the architecture suggested includes servers with RS 232 and RS 485 type series communications [SUN 14].

On the technical side, we highlight the following research. An article by Chaudhury *et al.* [CHA 15] notes that saturation of the radiofrequency spectrum has necessitated the invention of new technologies. Among these is LiFi able to reach high-speed Internet connections. LiFi, uses a broad frequency spectrum ranging from infrared to ultraviolet for communications that can theoretically reach 10 Gbps. LiFi is not confined only to LED diodes and laser technology. LiFi is a context that can find new paths for the present and future.

A state of the art on new challenges in visible light communication has been provided by Ergul. Connections with digital communications are suggested [ERG 15].

New research directions have been tackled in the case of LEDs, mainly for thermal management, by Karlicek [KAR 13].

The receivers' heterogeneity in visible communication has been studied by Hermans [HER 14]. It is a significant challenge as the receivers differ greatly in resolution, which results in variability in channel quality. A communication schema in which each receiver can decode a type of data has been suggested in Figure 7.2.

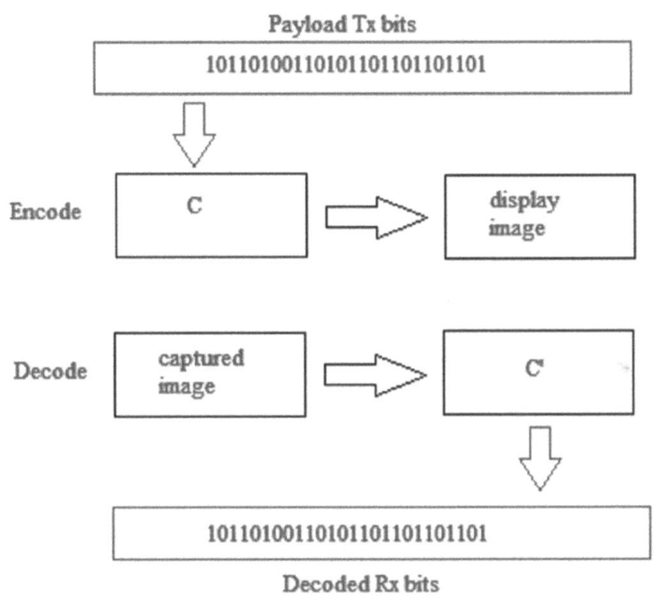

Figure 7.2. *Communication schema for the decoder*

Among LiFi applications, we will discuss the following two broad categories:

– applications in digital communications;

– applications in cryptography.

Dorward's article examines the economic impact of using highly integrated electronic command for the next generation of multichannel transmitter–receiver optical multichannels for access networks. Problems with cost due to the size and energy consumption are summarized, including the extensive costs that appear in the process of global manufacturing and user-friendliness for final products [DOR 12]. These elements of sizing and energy cost are fundamental for access networks.

Local LAN networks and in particular wireless networks have been used significantly and have promoted a myriad of wireless communications services and applications for ubiquitous computing.

7.2. Applications in digital communications: estimating the channel in OFDM systems

Compressed detection techniques make it possible to use the sparsity of radar scenes to potentially improve the system's performance. In this section, tools for compressed sensing are applied to MIMO radars to reconstruct the scene in the azimuth range Doppler domain. The conditions are derived from radar wave forms and transmit and receive it from boards such that the detection radar matrix has little coherence and rare recuperations become possible. Theoretical performance limits are shown and checked by digital simulations.

Recent studies have focused on the use of multiplexing by orthogonal frequency distribution (OFDM) with carrier equalization only in the frequency domain. According to Elgala [ELG 15], this technique combined with direct intensity modulation and direct detection can be used in LiFi. This has led us to discuss the following work, which can serve as a direct application of MIMO–OFDM in LiFi.

It is well known in all wireless communication systems that weak and coherent data detection is only possible if a precise estimation of the channel is available at the receiver. As a general rule, techniques for estimating channels for coherent OFDM modulations can be classed in two different families: techniques based on the transmission of training sequences (pilots) and blind estimation methods. Blind methods are based on the statistics of unknown data symbols and the channel's statistical properties and do not require a pilot symbol. In this chapter, we study methods of estimating the channel based on the training pilots for OFDM systems. We still however present an overview of the different approaches for estimating the channel in OFDM systems.

7.2.1. *Introduction*

Channel estimation can be carried out in different ways: with or without the help of a parametric model, using frequency and/or time correlation properties in the wireless channel, blind methods or those based on training pilots, adaptive or non-adaptive methods. Nonparametric methods try to estimate the quantities of interest (for example the frequency response) without relying on a model of a specific channel. Conversely, parametric estimation assumes a model with a certain channel, determines this model's parameters and from this deduces the quantities of interest. Time-correlated correlations and frequency spacing correlations are specific properties of the channel that can be incorporated in the estimation method in order to improve the quality of the estimation.

Estimation methods based on training pilots are the most widely used methods; they are applicable in systems where the transmitter sends a certain, known signal. Blind estimation, on the other hand, relies on some of the signal's properties (for example the signal's cyclostationarity) and is rarely used in practical OFDM systems. Adaptive estimation methods are generally used for channels with rapid time variation.

Estimation of the channel based on the training pilots can be achieved either by inserting pilot tones in all the OFDM symbols' subcarriers with a specific period or by inserting pilot tones in each OFDM symbol [DEL 09]. The first class is called channel estimation based on block pilots and has been developed under the hypothesis of a channel with slow fading. This type of pilot arrangement runs well when the channel's transfer function does not change very rapidly. The second class, an arrangement of comb pilots, can easily be used for monitoring fast channels. In comb arrangements, each OFDM symbol has some pilot tones; for this, it performs better in very variable environments.

There are two main problems in designing channel estimators for wireless OFDM systems. The first problem is the availability of data pilots, where the pilot means the reference signal used by the transmitter and the receiver. The second problem is designing an estimator of low complexity and good channel monitoring capacity. The two problems are linked. In general, the channel of the altered OFDM system can be considered as a bidimensional signal (2D) (time and frequency). The optimal channel estimator in terms of average quadratic error is based on 2D Wiener filter interpolation. Unfortunately, such a structure for the 2D estimator is very complex for practical implementation. The combination of the high data rate and the low binary error rate (BER) in the OFDM systems requires the use of estimators that have both low complexity and high precision, where the two constraints have opposite effects, hence the need for a good compromise. Single dimension (1D) estimators are generally adopted in OFDM systems to achieve the compromise between complexity and precision [FAZ 03, ETS 04, FOZ 05, LI 02, DOW 03, DEL 09, LEE 04]. The two basic 1D estimators are the estimators based on block-type pilots and the comb estimator. Block-type estimators can be based on the least squares (LS) method, minimal mean squared error (MMSE) and the modified MMSE. Estimators based on comb-type pilots include the LS estimator with 1D interpolation, the maximum likelihood estimator (MLE), and the parametric channel modeling based (PCMB) estimator. Other strategies for channel estimation have also been studied [COL 02, SHE 08, BAG 10, BEE 95, MUQ 99], such as estimators based on simplified 2D interpolations, estimators based on filtering and iterative decoding, estimators for OFDM systems with multiple transmission and reception antennae and so on.

7.2.2. *Interpreting time–frequency and pilot distribution*

The OFDM transmission system can also be described using a two-dimensional trellis in the time–frequency domain. Figure 7.3 shows such a representation in the time–frequency domain in which the OFDM symbols are transmitted sequentially. As Figure 7.3 shows, an OFDM symbol consists of N_c subcarriers, arranged symmetrically around the carrier frequency (or DC). It should be noted that only N_0 subcarriers on N_c are used for data transmission; no data are transmitted on DC and the subcarriers are at the extremities (to avoid interference in the adjacent bands).

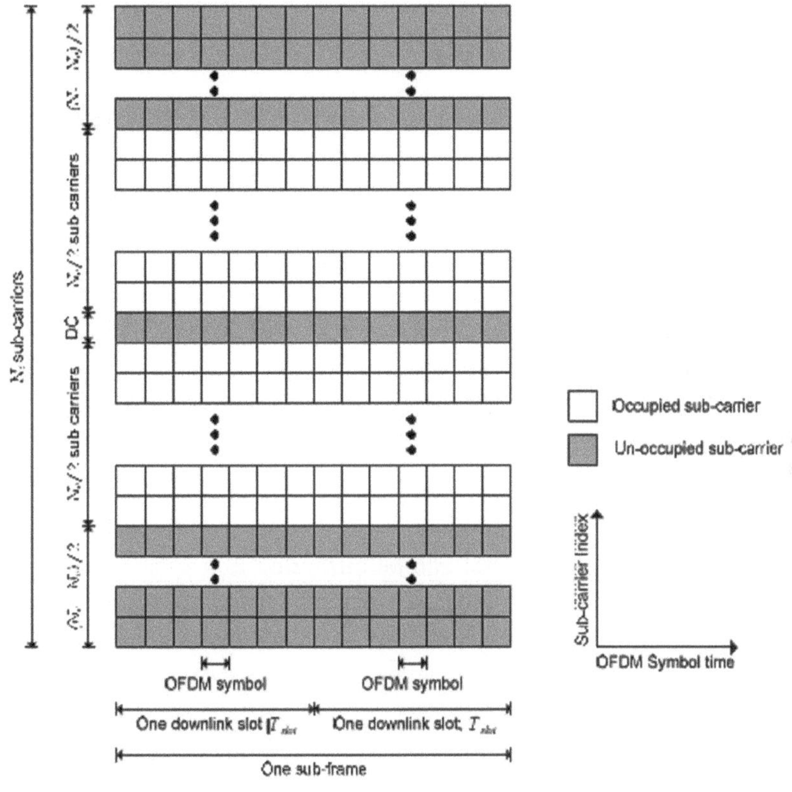

Figure 7.3. *Time–frequency trellis for OFDM transmission*

There are different possibilities for distributing pilots in the time–frequency domain of an OFDM system. We will discuss three of these possibilities as illustrated in Figure 7.4.

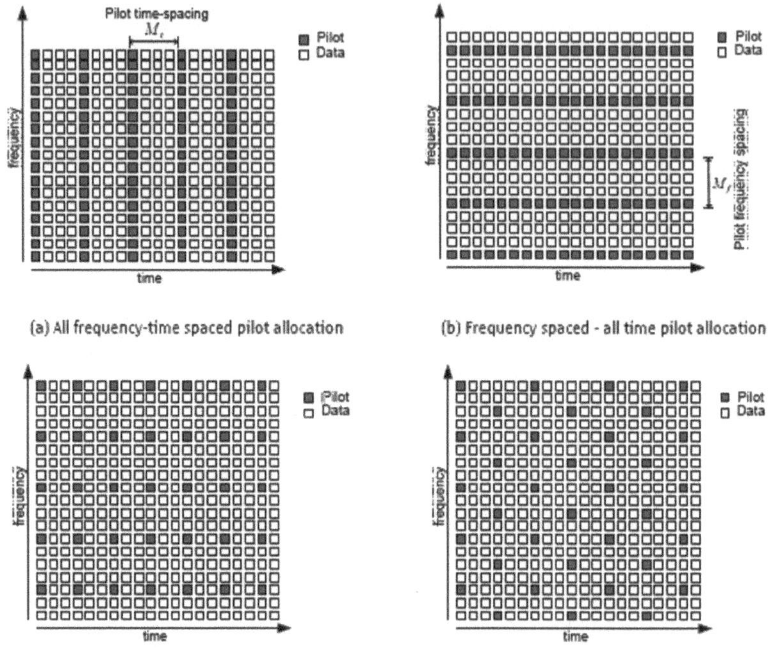

(a) All frequency-time spaced pilot allocation

(b) Frequency spaced - all time pilot allocation

(c) Frequency spaced - time spaced pilot allocation

(d) Frequency spaced - time spaced pilot allocation

Figure 7.4. *Different possibilities for distributing pilots*

An entire OFDM symbol can be attributed to a pilot, as indicated in Figure 7.4(a). Such an allocation will be very helpful for estimating the channels with highly dispersive frequency and channels with low Doppler effect to the detriment of the data rate.

The pilots can be transmitted on the individual subtransporters for the entire transmission period as shown in Figure 7.4(b). Such a strategy will be very advantageous for channels with selective moderated frequency and channels with high Doppler effect [GAL 10].

Training pilots can be attributed to intervals spaced in time and frequency as illustrated in Figures 7.4(c) and (d) depending on the spacing of the pilots in the time–frequency and channel's properties. Such a distribution strategy gives good results in channels with high selective frequency and channels with high Doppler effect.

Note that the frequency spacing M_f and the time spacing M_t between the pilots should be fairly reasonable for fully estimating the channel. For an ideal reconstruction, the frequency spacing M_f between the pilots should meet the conditions of the 2D sampling theorem [NEC 04]:

$$M_f < \frac{N_c}{2W\tau_{max}} \quad [7.1]$$

where:

– W is the width of the OFDM system's band;

– $\tau_{\max}$ is the channel's maximum excess delay;

– N_c is the number of subcarriers in the OFDM system.

Similarly, the time spacing between pilots M_t should meet the condition

$$M_t < \frac{W}{2(N_c+WT_{cp})(f_d+v)} \quad [7.2]$$

where:

– T_{cp} is the length of the cyclic prefix;

– f_d and v are the maximal Doppler propagation and the Doppler shift, respectively.

7.2.3. *Estimating the block-type channel*

In estimation based on block-type pilots, as Figure 7.5 shows, the OFDM symbols are transmitted periodically, and all the subcarriers are used as pilots. The task here is to estimate the channel conditions, taking account of the pilot signals and the signals received, with or without using specific knowledge of the channel's statistics. The receiver uses the channel conditions estimated to decode the data received inside the block until the following pilot symbol arrives. If the channel is constant throughout the block, there will be no estimation error, since the pilots are sent to all the carriers. Estimation can be based on LS, minimal mean quadratic error (MMSE) or the modified MMSE.

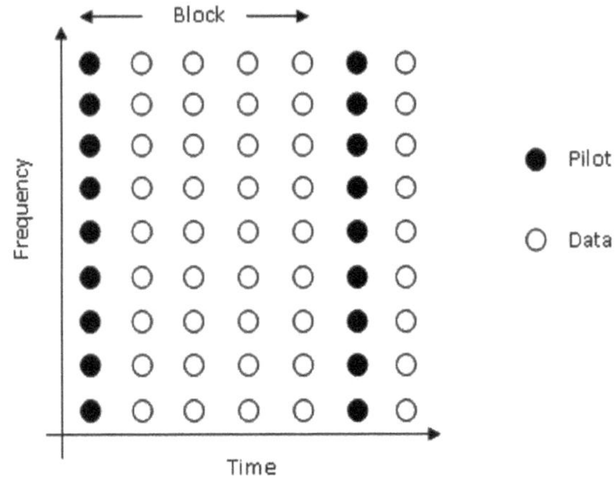

Figure 7.5. *Estimating the block-type channel*

This type of pilot arrangement is particularly adapted for slow fading radio channels, as the training block contains all the pilots. It is not necessary to interpolate the channel into the frequency domain; as a result, this type of pilot arrangement is relatively insensitive to frequency selectivity.

7.2.4. *Estimating the channel by LS*

We also define $H = DFT_N(g) = Fg$ and $W = Fw$.

Assuming that the interference is completely eliminated, we can obtain:

$$y = DFT_N(IDFT_N(X) \otimes \frac{g}{\sqrt{N}} + W = XFg + W = XH + W \quad [7.3]$$

The LS estimator minimizes the parameter $(Y\text{-}XH)^H(Y\text{-}XH)$, where $(.)^H$ designates the conjugate transpose operation. The LS estimate of H, taking account of the data received Y and the symbols sent X, is given as [BEE 95]:

$$\hat{H}_{LS} = X^{-1}Y = \left[\left(\frac{X_k}{Y_k}\right)\right]^T \quad ,k = 1, \ldots., N \qquad [7.4]$$

Without using knowledge of the channel's statistics, the LS estimator is calculated with low complexity, but it suffers from a substantial mean quadratic error.

7.2.5. *Estimating the channel by minimal mean quadratic error*

Estimation by MMSE has been demonstrated to be better than LS estimation for estimating the channels of OFDM systems based on the block-type pilot arrangement [DEL 09]. The main drawback of MMSE estimation is its great complexity, which increases exponentially with the observation samples [PET 04].

The MMSE estimator is the best linear estimator for mean quadratic error. It uses second-order statistics of the channel's conditions to minimize the mean quadratic error.

By R_{gg}, R_{HH} and R_{YY}, we mean the auto covariance matrices of g, H and Y, respectively, and by R_{gY} the cross-covariance matrix between g and Y. By σ_N^2, we also mean the variance of the noise $E\{|W|^2\}$. Let us assume that the vector of the channel g and the noise vector W are not correlated, then:

$$R_{HH} = E\,\{HH^H\} = E\{(Fg)(Fg)^H\} = FRggF^H \qquad [7.5]$$

$$R_{gY} = E\{gY^H\} = E\{g(XFg + W)^H\} = R_{gg}F^HX^H \qquad [7.6]$$

$$R_{yy} = E\{YY^H\} = XFR_{gg}F^HX^H + \sigma_N^2 I_N \qquad [7.7]$$

We assume that R_{gg} (therefore R_{HH}) and σ_N^2 are known at the receiver in advance; the MMSE estimator of g is given by [11]:

$$\hat{g}_{MMSE} = R_{gY} R_{YY}^{-1} Y^{HH} \qquad [7.8]$$

Note that if g is not Gaussian, then $\hat{g}_{MMSE}$ is not necessarily an estimator of minimal mean quadratic error, but it is always the best linear estimator in the sense of mean quadratic error. The MMSE estimator is given by:

$$\hat{H}_{MMSE} = F\hat{g}_{MMSE} = F\left[(F^H X^H)^{-1} R_{gg}^{-1} \sigma_N^2 + XF\right]^{-1} Y$$

$$\hat{H}_{MMSE} = F\hat{g}_{MMSE} = F\left[(F^H X^H)^{-1} R_{gg}^{-1} \sigma_N^2 + XF\right]^{-1} Y$$

$$= F R_{gg}\left[(F^H X^H X F)^{-1} \sigma_N^2 + R_{gg}\right] F^{-1} \widehat{H_{LS}}$$

$$= R_{HH}[R_{HH} + \sigma_N^2 (XX^H)^{-1}]^{-1} \hat{H}_{LS} \qquad [7.9]$$

7.2.6. *Estimating the comb-type channel*

The second type of pilot arrangement is shown in Figure 7.6. The pilot allocations are distributed uniformly in each OFDM block. Assuming that the loads on the pilot allocations are the same, the comb pilot arrangement has a greater rate of retransmission. Therefore, this system presents better resistance to channels with fast fading. Given that only some subcarriers contain the pilot signal, the channel's response to non-pilot subcarriers will be estimated by interpolating the adjacent pilot subchannels; hence, the conclusion is that the comb pilot arrangement is sensitive to frequency selectivity in comparison with the block pilot arrangement.

Estimation of the comb-type pilot tone has been introduced to meet the need for equalization during channel changes, indeed, in the same OFDM block. Estimation of the comb-type pilot channel consists of a set of algorithms for estimating the channel at the pilot's frequencies, and then an interpolation in the channel.

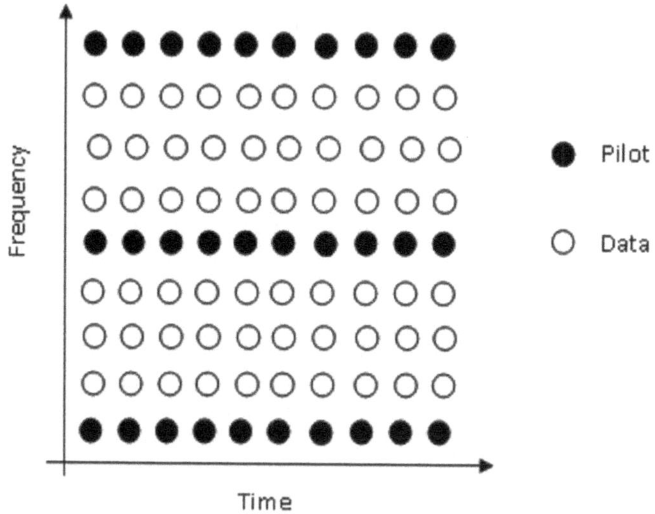

Figure 7.6. *Estimating a comb-type channel*

In channel estimation based on comb-type pilots, and as Figure 7.4 demonstrates, for each symbol transmitted, N_p pilot signals are inserted uniformly in X with S different subcarriers, where $S = N/N_p$.

The receiver knows the locations of pilots $P = [P_k]^T$ $(k = 1, \ldots, N_p)$, the values of pilots: $X_p = [X_k^p]^T$ $(k = 1, \ldots, N_p)$, and the signal received Y. LS estimations of the conditions of the channel at the subcarriers (H_p) are calculated by:

$$\widehat{H}_{LS}^{p} = \left[\frac{X(P_1)}{X_1^p}, \frac{X(P_{21})}{X_2^p}, \ldots, \frac{X(P_{Np})}{X_{Np}^p}\right]^T \qquad [7.10]$$

At this point, the task is to estimate the channel's conditions at the level of the subcarriers' data (described by H with a length N), given that the LS estimates at the pilot's subcarriers are $\widehat{H}_{LS}^{P}$, the signal received Y, and there can be some additional statistical data on the channel. Solutions include the LS estimator with a 1D interpolation, the MLE, and the PCMB estimator.

7.2.7. *LS estimator with 1D interpolation*

Linear interpolation (LI) is the simplest method for estimating a channel from the vector of the channel's estimates at the pilot's frequencies. This is done by LI of the vector of the channel's estimates for the pilot's two closest subcarriers. However, LI presents limited noise reduction for the channel's estimates for the localization data. In fact, it is the solution's simplicity that also makes it attractive. An additional gain can be obtained by averaging the channel's estimates using a sliding window. It can be noted that the length of the window (number of pilots used in the window) is inversely proportional to the coherence of the channel's bandwidth. For a channel with a high-frequency coherence bandwidth (a small excess delay), a large number of the channel's interpolated coefficients can be used to make the average and vice versa.

After estimating the transfer function of the pilot tones, the transpose of the data pilot channel can be interpolated according to the adjacent pilot tones. LI has been tackled in [MAH 10], and it has been demonstrated that it is better than interpolation by point distribution.

7.2.7.1. *Linear interpolation*

In the LI algorithm, two of the pilot's successive subcarriers are used to find the channel's response for data from the subcarriers that are situated between the pilots. This is given by [BAG 10]:

$$\hat{H}(ks+t) = \hat{H}^p_{LS}(k) + \left(\hat{H}^p_{LS}(k) - \hat{H}^p_{LS}(k)\right)\frac{t}{s} \quad 0 \leq t \leq s \qquad [7.11]$$

LI of the channel can, therefore, be implemented through a digital filter such as the Farrow structure [DEL 09].

7.2.7.2. *Cubic spline interpolation*

Cubic and spline interpolations can be made using the "spline" function in MATLAB. Cubic spline interpolation (SCI) gives a continuous, regular polynomial that is adequate for the data of the points shown. Spline interpolations are more efficient than LI for comb-type pilot arrangements [DOW 03].

7.2.7.3. *Low-pass interpolation*

The low-pass interpolation (LPI) method is carried out by inserting zeros into the original sequence $\widehat{H}_{LS}^{P}$, and then applying a low-pass filter with finite impulse response (the "interp" function in MATLAB), which makes it possible for the original data to pass unchanged. This method also interpolates in such a way that the mean quadratic error between the interpolated points and their ideal values is also minimized.

7.2.7.4. *Time domain interpolation*

The time domain interpolation (TDI) method is a high-resolution interpolation based on zero-padding and DFT/IDFT (Discrete Fourier Transform and Inverse discrete Fourier Transform). It first converts $\widehat{H}_{LS}^{P}$ to the time domain using the IDFT, then interpolates the sequence of N points from the time domain with the interpolation point distribution method [KAT 02]. Finally, the DFT converts the interpolated sequence from the time domain into the frequency domain.

7.2.7.5. *Second-order interpolation*

The second-order interpolation (SOI) method is more effective than the LI method, where estimation of the channel at the data subcarrier is obtained by a linear weighted combination of the three estimations of the adjacent pilots.

In [KAT 02], the performance of these estimation techniques is classified from best to worst as follows: LPI, SCI, TDI, SOI and LI. Moreover, LPI and SCI interpolation give almost the best performance even in scenarios with weak and average signal to noise ratio (SNR), whereas LPI interpolation is better than SCI interpolation for high SNRs. In terms of complexity, TDI, LPI and SCI interpolations are almost as difficult as calculations, whereas SOI and LI interpolations are less complex. Consequently, LPI and SCI interpolations are generally recommended, as they present the best compromise between performance and complexity.

7.2.8. *Maximum likelihood estimator*

We define $g_{L+1} = [g0, ..., g_{L+1}]^T$ the first $(L + 1)$ coefficients of g, and the non-square DFT matrix:

$$F_{A,B} = \left[W_N^{a,b}\right]_{AxB} \quad 0 \leq a < A, 0 \leq b < B \qquad [7.12]$$

We also define the uniform line-spacing DFT matrix S as follows:

$$F(S)_{A,B} = \left[W_N^{aS,b}\right]_{AxB} = \left[W_N^{a,bS}\right]_{AxB} \quad 0 \leq a < A, 0 \leq b < B \qquad [7.13]$$

It is clear that $H^p = F(S)_{Np\times(L+1)}g_{L+1}$, where S is the spacing between the pilots' subcarriers. The MLE of g_{L+1}, given the estimates of Hp (H^p_{LS} is used) is then obtained by:

$$g_{L+1} = \left(F(S)^H_{Np,L+1} x F(S)_{Np,L+1}\right)^{-1}\left(F(S)^H_{Np,L+1}\right)\widehat{H}^p_{LS} \qquad [7.14]$$

Finally, complete estimation of the channel $\widehat{H}$ of all the subcarriers is calculated from g_{L+1} by:

$$\widehat{H}_{MLE} = F_{N,L+1}\widehat{g}_{L+1} \qquad [7.15]$$

7.2.9. *PCMB estimator*

The channel is modeled by multipath fading with M resolvable paths with a complex path gain $\{\alpha_m\}$ and time delays $\{\tau_m T_s\}$. We assume that the different path gains are not correlated with one another. By $R_\alpha(M)$, we mean the channel's autocovariance matrix, and $R\alpha(M) = diag\{\sigma\alpha 1, ..., \sigma\alpha M\}$. In [LEE 04], a system for estimating the channel based on the parametric modeling of the channel's parametric modeling is suggested. In this estimator, knowledge of the channel is needed, i.e. M and $\{\tau_m T_s\}$ are needed. Estimation of M is obtained by the criteria of the minimal description length (MDL). Estimation of a signal's parameters by the rotational invariance method (ESPRIT) [COL 02] is used to acquire the initial time delays for multiple paths, and an interpath interference cancellation (IPIC) with a delay locked loop (DLL) to provide monitoring for the time delays in the multipath channel. Two non-uniform interspace DFT matrices are defined as follows:

$$B_{Np,\widehat{M}} = \left[W_N^{p(k)\hat{\tau}(m)}\right]_{Npx\widehat{M}} \quad 0 \leq k < 1, 0 \leq m \leq \widehat{M} \qquad [7.16]$$

$$B_{N,\widehat{M}} = \left[W_N^{i\hat{\tau}(m)}\right]_{Nx\widehat{M}} \quad 0 \leq i < 1, 0 \leq m \leq \widehat{M} \qquad [7.17]$$

where $\{P(k)\}$ are the pilots' positions and $\hat{\tau}\{(m)\}$ are the estimated time delays in the multipath channel.

The MMSE estimator is given by:

$$\widehat{H}_{PCMB} = B_{Np,\widehat{M}} \left(\frac{\gamma}{\overline{SNR}} R_{\alpha}(\widehat{M})^{-1} + B_{Np,\widehat{M}}^{H} B_{Np,\widehat{M}}\right)^{-1} B_{Np,\widehat{M}}^{H} \widehat{H}_{LS}^{p} \qquad [7.18]$$

where $\overline{SNR}$ is the mean of the SNR and γ is the ratio of the signal's mean power to the pilot's power.

If we compare these three estimation methods, the LS with 1D estimator interpolation shows a more reduced complexity. Without counting the complexity of the MDL and IPIC–DLL estimator for the following channel's parameters, the PCMB estimator is generally less complex than the ML estimator if $2\widehat{M} < N_P$ [RAE 99]. Moreover, the LS estimator with 1D interpolation is less effective than the two others in terms of MSE (Mean Squared Error) and RCS (Radar Cross Section). The performance of the PCMB estimator and the ML estimator is almost the same, even though the PCMB estimator gives better results in terms of MSE for small SNR values.

There are other estimation methods, such as simplified 2D estimators, iterative estimators and channel estimators, for OFDM systems with multiple transmission and reception antennae.

7.3. An application in cryptography: the SAC–OCDMA system

7.3.1. *Introduction*

For data traffic densification on telecommunications networks, multiple access techniques, frequency division multiple access and time division multiple access, have been developed. These multiplexing processes are used in RF transmissions, as well as in optical communication systems.

Wavelength division multiplexing is a transposition of FDMA (Frequency division multiple access) to optical networks. CDMA Code division multiple access (CDMA) is a multiplexing system defined as a spread spectrum [GLI 97]. This technique was initially designed for military applications. It makes it possible, by spreading power over a broad range of channel frequencies, to have a better resistance to the phenomenon of fading and gives the signal transmitted the form of noise, which makes it difficult for receivers for which it is not intended to detect.

Several studies on the transposition of CDMA techniques to optical communication systems date from the last 20 years. Taking advantage of the large bandwidth available in an optical channel, the optical CDMA technique (OCDMA) also aims to increase the capacity of multiplexing by increasing the number of users at the cost of deteriorating the quality of the link [PRU 86, MAR 93].

We can distinguish two approaches to OCDMA; the coherent and incoherent approaches. In the incoherent OCDMA system, multiple access interference (MAI) is one of the main limitations resulting from the use of unipolar codes. To reduce this effect [ANU 09], several techniques are suggested to eliminate the contribution from unwanted users. The technique of multiple accesses by code distribution in the spectral domain (spectral amplitude cordage (SAC)–OCDMA) gives a better solution for reducing the effect of interference from the code used with flexible cross-correlation (FCC) [WEI 01].

The literature contains many studies that aim to develop new codes for the SAC–OCDMA system such as orthogonal optical codes

(OOC) [WEN 02, DJE 16], the "Prime" code [WEN 02], the frequency-hopping "MFH" [ANU 06], the double-weight (DW) [ANU 07], the modified double-weight (MDW) [WEI 02], the random diagonal (RD) [ALJ 05] and the FCC [ALJ 12]. However, these codes suffer from various limitations such as complications in construction, low cross-correlation and codes lengths that are too long.

In this section, we suggest a new code family: the SAC–OCDMA. Since the concept of the code is based on FCC, we give it the name modified flexible cross-correlation (MFCC).

7.3.2. *The concept of code*

7.3.2.1. *Mathematical preliminaries*

Optical codes belong to the family of K (K users) binary sequences [0, 1], the length N, a maximal cross-correlation, λ_{max} and a weight w. The optimal code set is one that has minimum cross-correlation properties to support a maximal number of users with a minimal code length [HIL 09]. This guarantees a better quality of service with a low probability of error for a given number of users, at least for short-distance optical networks.

It has been shown that the main bottleneck in implementing optical networks is basically the MAI when all users try to send data simultaneously. This problem can be overcome by designing sequences of code such that they can at least cause an overlap between data chips [RAS 13].

If A = {an} and B = {bn} are of such length that:

$$\{a_i\} = '0' \quad or \quad '1'. \qquad i = 0, \ldots\ldots, N-1$$

$$\{b_i\} = '0' \quad or \quad '1'. \qquad i = 0, \ldots, N-1 \qquad [7.19]$$

The autocorrelation and cross-correlation functions of these sequences are defined by:

$$\lambda_a(\tau) = \sum_{n=0}^{N-1} a_i a_{i+\tau} \quad [7.20]$$

$$\lambda_{ab}(\tau) = \sum_{n=0}^{N-1} a_i b_{i+\tau} \quad [7.21]$$

where "an" is a binary sequence {0, 1}, the maximal value of λa (τ) in equation [7.20] is for τ = 0 and is equal to w, the sequence's weight. Therefore,

$$\lambda_a(0) = w \quad [7.23]$$

If λ_{xm} and λ_{xym} indicate the maximal autocorrelation and cross-correlation values, respectively, then an optical code of length "N" and weight "W" can be written as (N, W, λ_{xm}, λ_{xym}) or X_{max} = max {λ_{xm}, λ_{xym}}. It can also be noted that for an optical code with a weight "W", the autocorrelation can be written as:

$$\lambda_{a\ \max} = \sum_{n=0}^{N-1} a_i a_i = w \quad [7.24]$$

In practice, for K users it is necessary to have a number K of codes in a set of given values of N, W, λ_{am} and λ_{abm}. The codes described by equation [7.19] can also be represented in the form of a vector as:

$$A = \{a_i\} \quad for \qquad i = 0, \ldots\ldots, N-1 \quad [7.25]$$

$$B = \{b_i\} \quad for \qquad i = 0, \ldots, N-1 \quad [7.26]$$

where A and B are vectors of length "N" with elements as defined by equation [7.26]. In terms of vectors A and B, equation [7.20] and equation [7.21] are written as follows:

$$\lambda_A(0) = AA^T = w \quad [7.27]$$

7.3.2.2. *Constructing code*

The MFCC code is designed using the following three parameters: N, w and λ_c that, respectively, represent the length of the code, the weight of the code (the number "1" in each word of the code) and the cross-correlation of the suggested MFCC code. The new optical MFCC code is represented in a matrix K × N, where K lines represent the number of users and N columns represent the minimal length of the code. It is formed of two submatrices A and B, as follows:

$$\mathrm{MFCC} = [\mathrm{A}||\mathrm{B}]_{\mathrm{K}\times\mathrm{N}} \tag{7.28}$$

Some properties can be noted during construction of the suggested code, including:

– the first submatrix A is designed such that cross-correlation (λ_c) between two consecutive cubes is equal to 0, although the matrix B is designed with $\lambda_c = 1$;

– the length of the matrix should be equal to K, where K represents the number of users, whereas the length of matrix B should be equal to K + 1; thus, the total length of the MFCC code is equal to:

$$\mathrm{N} = 2(\mathrm{K}-1) + \mathrm{w} \tag{7.29}$$

The autocorrelation and cross-correlation property can be written as:

$$\sum_{i=1}^{N} C_k(i)C_l(i) = \begin{cases} w; & k = l \\ 1\ ; & k \neq l \end{cases} \tag{7.30}$$

where $C_k(i)$ represents the second component of the Kièm MFCC code.

Using the properties mentioned above, the MFCC code for three users can be written as follows:

$$MFCC = \begin{bmatrix} 1 & 0 & 0 & 1 & 1 & 0 & 0 \\ 0 & 1 & 0 & 0 & 1 & 1 & 0 \\ 0 & 0 & 1 & 0 & 0 & 1 & 1 \end{bmatrix} \qquad [7.31]$$

in this case, $K = 3$ and $N = 7$ for $w = 3$.

The code word for each user according to [7.31] becomes:

$$MFCC = \begin{cases} User\ 1; & \lambda_1, \lambda_4, \lambda_5 \\ User\ 2; & \lambda_2, \lambda_5, \lambda_6 \\ User\ 3; & \lambda_3, \lambda_6, \lambda_7 \end{cases} \qquad [7.32]$$

7.3.3. *Digital analysis of the MFCC code's performances*

To evaluate the suggested MFCC code's performances, the SNR and the binary rate error (BER) should be calculated using the following equations:

$$SNR = I^2 / \sigma^2 \qquad [7.33]$$

where I represents the incident current, expressed by:

$$I = \Re \int_0^\infty G(v)dv \qquad [7.34]$$

and σ^2 is the noise variance, which is the sum of the powers of the "Shot", phase induced intensity noise (PIIN) and thermal noise:

$$\sigma^2 = I^2_{shot} + I^2_{PIIN} + I^2_{thermal} \qquad [7.35]$$

From equation [7.34], $\Re$ represents the photodiode PIN.

Moreover, the power spectral density is given from [WON 03] by:

$$G(v)=\sum_{k=1}^{k}d_k\begin{pmatrix}\sum_{i=1}^{N}C_k(i)C_l(i)\\-\sum_{i=1}^{N}C_k(i)\left(C_k(i)C_l(i)\right)\end{pmatrix}\left\{u\frac{\Delta v}{N}\right\}\qquad[7.36]$$

where

$$\sum_{i=1}^{N}C_k(i)C_l(i)=\begin{cases}W; & k=l\\1; & k\neq l\end{cases}\qquad[7.37]$$

$$\sum_{i=1}^{N}C_k(i)\left(C_k(i)C_l(i)\right)=\begin{cases}1; & k=l\\1; & k\neq l\end{cases}\qquad[7.38]$$

$$\begin{aligned}&\sum_{i=1}^{N}C_k(i)C_l(i)-\sum_{i=1}^{N}C_k(i)\left(C_k(i)C_l(i)\right)\\&=\begin{cases}W-1; & k=l\\0; & k\neq l\end{cases}\end{aligned}\qquad[7.39]$$

From equations [7.34] and [7.37], the incident current becomes:

$$I=\Re\frac{p_{sr}}{\Delta v}(W-1)\qquad[7.40]$$

and the noise powers are given as follows:

$$I_{shot}^2=2eBI=2eB\Re\int_0^{\infty}G(v)dv\qquad[7.41]$$

$$I_{shot}^2=2eB\Re\frac{p_{sr}}{\Delta v}(W+3)\qquad[7.42]$$

$$I_{PIIN}^2=BI^2\tau_c=B\int_0^{\infty}G(v)^2dv\qquad[7.43]$$

$$I_{PIIN}^2=\frac{B\Re^2P_{sr}^2KW}{N^2\Delta v}(W+3)\qquad[7.44]$$

and

$$I^2_{thermal} = \frac{4K_b T_n B_r}{R_L} \qquad [7.45]$$

From the above-mentioned equations, the SNR can be described as follows:

$$SNR = \frac{\left(\frac{\Re P_{sr}(W-1)}{N}\right)^2}{\left[\frac{2eB_r \Re P_{sr}}{N}(W+3) + \frac{B\Re^2 P_{sr}^2 KW}{N^2 \Delta v}(W+3) + \frac{4K_b T_n B_r}{R_L}\right]} \qquad [7.46]$$

All the parameters of equation [7.46] are described in Table 7.1.

Parameter	Description
E	Electron charge
B	Passband
Kb	Boltzmann constant
Tn	Absolute temperature
RL	Load resistance

Table 7.1. *Description of parameters*

The BER is obtained using the SNR by the following equation:

$$BER = \frac{1}{2} erfc\left(\sqrt{\frac{SNR}{8}}\right) \qquad [7.47]$$

7.3.4. Results and discussions

The system's performances are analyzed using the parameters listed in Table 7.2 and the digital results are shown in Figures 7.7–7.9.

To compare our code's performances with those of existing codes, we have traced the BER as a function of the number of users for our code and two other codes; the "MDW" code and the "RD" code with a weight W = 4. It can be seen that our code offers improved performances compared to the MDW and RD codes, with a low BER (less than 10^{-9}).

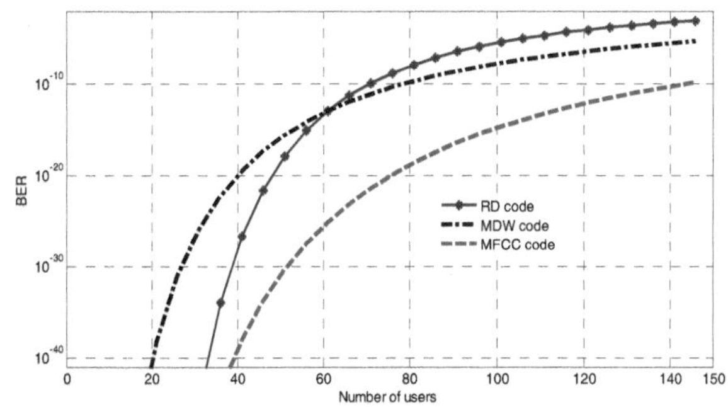

Figure 7.7. *BER variation depending on the number of users for MFCC, MDW and RD codes. For a color version of the figure, see www.iste.co.uk/benslama/networks.zip*

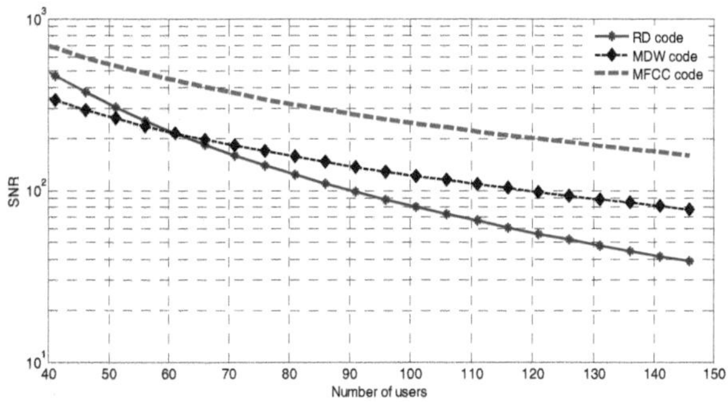

Figure 7.8. *SNR variation as a function of the number of users for the MFCC, MDW and RD codes. For a color version of the figure, see www.iste.co.uk/benslama/networks.zip*

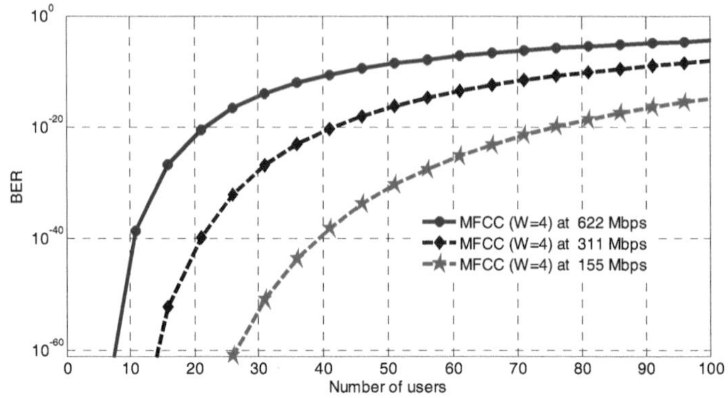

Figure 7.9. *Variation of the BER as a function of the number of users for the MFCC code with different data transmission rates. For a color version of the figure, see www.iste.co.uk/benslama/networks.zip*

Quantum effectiveness	$\eta = 0.75$
Spectral width of broadband light source	$\Delta\lambda = 30$ nm ($\Delta\lambda = 3.75$ THz)
Operating wavelength	$\lambda_0 = 1{,}550$ nm
Passband	B = 155 MHz
Data transmission rate	Rb = 622 Mbps
Receiver noise temperature	Tn = 300K
Load resistance	RL = 1030Ω
Boltzmann constant	$K_b = 1.38 \times 10^{-23}$ W/K/Hz
Electron charge	$e = 1.60217646 \times 10^{-19}$ coulombs
Light speed	$C = 3 \times 10^{8}$ m/s

Table 7.2. *Parameters used in digital calculations*

Figure 7.7 shows a comparison of the performance of MFCC codes with the MDW and RD codes in terms of the variation in the SNR and the number of users; it can be seen that the suggested MFCC code displays significantly better performances (with a higher SNR) than those of the other codes, which ensures a good quality of data transmission.

Finally, the effect of the data flow rate performances of the suggested MFCC code is studied, by tracing the “BER” for different values of Rb, as illustrated in Figure 7.10. From these curves, we conclude that there is an increase in the data flow rate and a degradation of the system's performances.

In Figure 7.10, we see the design between the length of the fiber and the system's BER performances for the code MFCC (W = 3) for different binary flows, 155 and 622 Mbps. We can see that the BER system's performance 10-9 MFCC (W = 3) 155 Mbps will have significantly better BER during the system's running where it is capable of reaching a BER, which is acceptable as the system's running performance until the fiber is 55 km. It is different when the MFCC (W = 3) is 622 Mbps, where it can only run up to a fiber length of 44 km with the same running system BER threshold. From this observation, MFCC (W = 3) to 155 Mbps can successfully eliminate and remove the effects of PIIN and MAI for the SAC–OCDMA encoding system.

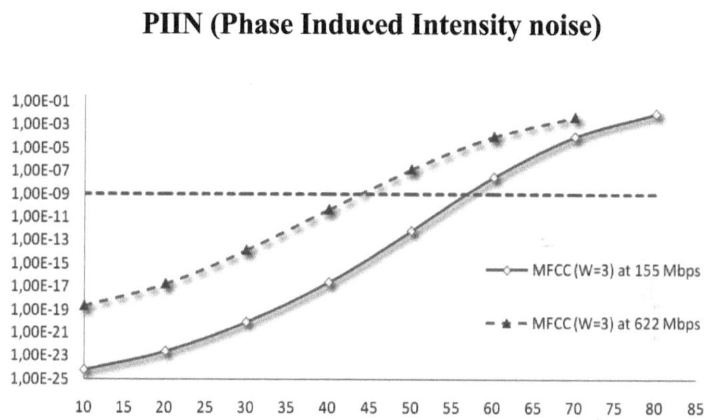

Figure 7.10. *BER Variation depending on fiber length for the MFCC code with different binary data flows (W = 3, K = 3). For a color version of the figure, see www.iste.co.uk/benslama/networks.zip*

Figure 7.11 shows the BER's variation depending on the different binary flows using the MFCC code. The flow varies from 155 Mbps to 3 Gbps with the same Hamming weight w = 3; the figure shows that with increase in the binary flow, BER will decrease. We can conclude that when we increase the binary flow, the pulse width decreases, thus producing sensitive bits that signal the dispersion effect.

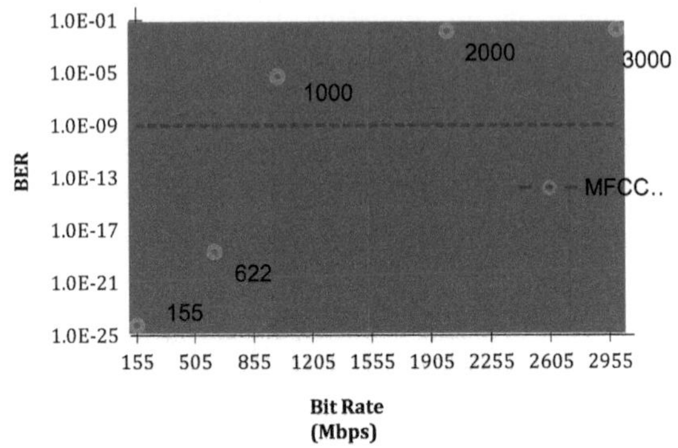

Figure 7.11. *BER versus fiber length at different bit rates MFCC (W = 3, K = 3). For a color version of the figure, see www.iste.co.uk/benslama/networks.zip*

Figure 7.12 shows that the BER will increase when there is an increase in the code's length. Each curve represents another weight value. It is clear that the BER is weaker for a large weight, w = 10, in comparison to a weight equal to 4, 6 and 8, for a constant number of users and a power equal to each spectral component. The high number of the code's length, without increasing the weight, means that a high passband number will be needed, which will produce noise. Moreover, the extra weight will make it possible to improve the BER due to the efficiency in reception.

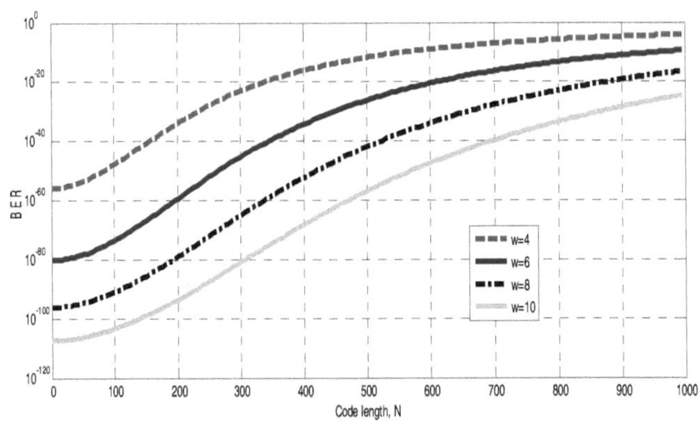

Figure 7.12. *MFCC code with different code length compared to the BER. For a color version of the figure, see www.iste.co.uk/benslama/networks.zip*

Figure 7.13 shows the power impact of the actual user on the BER at a code weight w = 4, 6, 8 and 10 for 30 users. The impact of the code weight is seen on a broad range of effective power. It is clear from this graph that the number of users attempting to reach the optical channel increases with a substantial effective power received at the receiver, the BER decreases significantly.

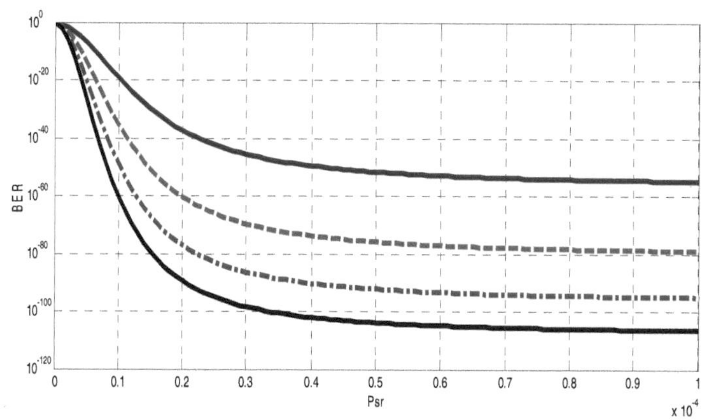

Figure 7.13. *Variation in the BER depending on the effective power for 30 users, w = 4, w = 6, w = 8 and w = 10. For a color version of the figure, see www.iste.co.uk/benslama/networks.zip*

7.6. *Conclusion*

A new FCC (MFCC) has been studied and put forward for the OCDMA optical transmission system. Analysis was carried out by calculating the SNR and the BER by taking account of the shot effect, PIIN (Phase Induced Intensity Noise) effect and sources of thermal noise. To evaluate the performance of the suggested code, the digital results obtained from implementing this code are compared to those of two other codes: MDW and RD.

From this comparison, it is seen that the MFCC code is the best in terms of BER compared to MDW and RD codes. Moreover, the simplicity in building this code and flexibility in monitoring cross-correlation makes this code a candidate for future applications based on the OCDMA technique.

8

Compressed Sensing in WiFi Systems

8.1. Preliminaries on compressed sensing in WiFi systems

The task of transmitting information from one point to another is a common and well-understood exercise. But the difficulty of efficient transmission or efficient information sharing between a large number of nodes remains a challenge, which results principally from the fact that we still have not developed the theories and tools for processing distributed signals, communications and information theory for large-scale network systems.

Several studies have been developed in the context of compressed sensing in wireless systems; we cite work by Tian and Giannakis [TIA 07], Chen and Wassel [CHE 12] and Zhang *et al.* [ZHA 12]. These two articles are quite full in terms of data. The sampling effect for better compression in wireless systems has been studied by Mahmoud *et al.* [MAH 10]. The energy factor, which is a basic element in the study of networks, has been studied by O'Connor *et al.* [OCO 14] and Fazel *et al.* [FAZ 11]. Luo [LUO 10] has shown that compressed sensing increases flow.

The problem can be illustrated by a simple example. We consider a network of n nodes, each with an information or data element x_j, $j = 1, \ldots, n$. These data may be files to be shared or simple scalar values corresponding to the allocated nodes or sensor measurements. To illustrate this example, suppose that each x_j is a scalar quantity. These collective data $x = [x_1, \ldots, x_n]$ are arranged along a vector; they are called network data to underline both the distributed nature of the data and the fact that they are shared across the network's adjacent communication infrastructure. The network data vector can be large: n can be a thousand, a million or more. Thus, even the process of collecting x in a single point is prohibitive, requiring at least n communications, yet this general meaning of the data network is crucial in applications ranging from network security to WiFi detection. Let us suppose, however, that it is possible to build a much-compressed version of x, efficiently and in a decentralized way. This would provide many advantages on condition that the compressed version can recuperate x with reasonable precision [QUE 09].

There are several decentralized compression strategies that might be useful. One possibility is that correlations between the data at different nodes are known initially. Source encoding techniques such as Slepian–Wolf encoding can be used to design compression schemas without collaboration between nodes [PRA 02]. Unfortunately, in many applications, knowledge of precise data correlations is not available, which makes it difficult, if not impossible, to apply any source encoding technique. This situation necessarily creates collaboration between processing and compression, or unknown correlations and dependencies between the network data can be used by exchanging information between the network's nodes [MOS 09]. However, design and implementation of collaborative processing algorithms can be very difficult as they rely on prior knowledge of anticipated correlations and depend on the nodes' processing capabilities [HAU 08].

The growing interest in indoor location services has created a need for greater precision in positioning in-house solutions in real time. The sparse nature of localization makes the theory of compressed sensing interesting for indoor positioning with the help of the power of the signal received from the WiFi access points (WLAN). To do this, a precise positioning system based on the power received from the signal has been suggested, using compressed sensing. This is a process for recuperating sparse signals from a small number of noisy measurements by solving a minimization problem of norm L_1. The localization estimator is formed of any localizer where the power of the signal received is compressed to a certain number of clusters to detect in which of the clusters the node is localized after a precise localization stage using compressed sensing to refine the localization. It has also been shown that compressed sensing can be used to reconstruct the radio map of the signal's power from measurements with only a small number of digital fingerprints, which considerably reduces the number of measurements. The suggested system is implemented on an integrated mobile device without WiFi and performances are evaluated. The test results obtained indicate that the proposed system leads to a substantial improvement in the precision and complexity of localization compared to methods widely used for classical digital fingerprints [WIG 14].

A significant observation is that the localization problem presents a sparsity inherent in the spatial domain. This is even clearer if we use a grid representation grill of the physical space where each cell in the grill indicates a position. We observe that the mobile's position is sparse on the ground. This prompted us to explore the theory of compressed sensing to reformulate the estimation of the positioning as a sparse approximation problem [NIK 11].

On the other hand, the power of the signal received from compressed sensing offers recuperation of the sparse signal from a small number of measurements. These measurements are obtained by solving a minimization problem of norm L_1. A pre-treatment orthogonalization procedure is used to induce the incoherence needed in the theory of sparse sensing. With the aim of mitigating the effects of variations in the power of the signal received due to the channel, the positioning system suggested is formed of two stages; a rough localization using affinity propagation, and exact localization using compressed sensing. In the exact localization stage, the problem of selecting the access point is studied to increase precision. To evaluate performance, an integrated WiFi mobile device is implemented [MAS 13].

The network constantly generates an enormous quantity of rich, diverse data. These data create interesting opportunities for network analysis. However, a major challenge for enabling effective analysis of the network is the presence of missing data and the measurement of errors and anomalies. Despite important work in network analysis, the main questions remain: (1) existing works do not take explicit account of anomalies or measurement noise, and encourage a serious degradation of performance through noise or significant anomalies, and (2) they suppose that network matrices have a low rank structure, which cannot hold in reality.

To answer these questions, in this chapter, we are developing LENS decomposition, a new technique for precisely decomposing a network matrix into a low rank matrix, a rare anomaly matrix, an error matrix and a small noise matrix. LENS has the following positive properties:

1) It is general: it can effectively support matrices with or without anomalies, with or without low rank.

2) Its parameters are self-regulating so that it can adapt to different types of data.

3) It is precise, as it integrates knowledge of the domain, such as the time location, spatial location and initial estimation (e.g. obtained from models).

4) It is polyvalent and can take charge of many applications, including those that lack value-interpolation, prediction and detection of anomalies.

We apply LENS to a broad range of network matrices – 3G, WiFi, mesh, sensor networks and the Internet. Our results show that LENS is clearly more defective than state-of-the-art compression of detection systems.

Practical results have also been obtained by [DUA 12], [SHI 08], [OKA 09] and [MAS 13].

8.2. Interconnection of WiMAX mobile and satellite DVB-S/RCS networks

8.2.1. *Introduction*

Telecommunications by satellite are the only system capable of covering substantial land surfaces with a high bandwidth without local topography being an issue. On the contrary, the architectural difference that exists between land satellite and mobile networks creates a problem of size in attempting a successful vertical handover (HO) between the two.

In this chapter, we discuss the different norms and standards that govern satellite communications. Then we devote particular attention to the DVB-RCS (digital video broadcast – return channel for satellite) norm, which concerns installation of a return channel using satellite for interactive services. With the aim of ensuring transparency in passage between WiMAX mobile and satellite networks combining DVB-S (digital video broadcast via satellite) and RCS systems, we focus our research on creating an interconnection model for these two.

With macro-mobility management at layer 3+ we carry out a comparative study focused on using the SCTP (*Stream Control Transmission Protocol*), then mobile SCTP (MSCTP) and cellular SCTP (CSCTP). This initiative involves defining the transport protocol, which suggests the most optimal Quality-of-Service (QoS) for a satellite WiMAX mobile connection.

8.2.2. *Satellite: a communication tool*

Whether it is created for the purpose of monitoring, calculating or observing, satellite remains a communication tool because it transmits the data that it collects. From there, the steps making it possible to transform it into communication relay were taken in 1962 with Telstar 1 or indeed relay [INT 03]. These satellites moving at low altitude quickly demonstrated the contribution of satellites for telecommunications, with the diffusion of television transmissions, but also with telephone links between the United States, Great Britain and France.

However, this method of moving has been scarcely used for television broadcasts. It was in 1965, with Early Bird, that a commercial telecommunication network using geostationary satellites,

Intelsat, was created, providing the diffusion of 36 simultaneous programs [LIA 99].

Among the four types of satellite (LEO, MEO, HEO and GEO), only the GEOs offer a relay fixed in relation to a point on Earth, suggesting coverage able to reach a third of the globe. Digital television therefore has its support medium of choice in such satellites: at the end of the 1990s, the MPEG-2 (Moving Picture Experts Group 2) [IZU 13] revolutionized the world of images and opened up the prospect of digital television. Therefore, the satellite had real success over time as a privileged support for this service.

The great diversity in the world of space telecommunications offers geostationary satellites very specific to each constructor and each service, posing a real problem with integrating IP onto these systems. In this context, the DVB norm seems called to be an important element. Ever more widespread, this norm is in the process of becoming a global standard, offering a shared base for satellite systems and aiming for a shared interface within stances of normalization such as ETSI.

8.2.3. *The DVB norm*

The DVB norm is a recent one, resulting from MPEG-2 norm. The last two appeared at the same time. The DVB standards were published in January 1995 by the European Broadcasting Union, and introduced basic techniques for installing digital TV transmission in Europe, Asia, Australia and many other regions. Its only competitor seems to be the American standard HDTV (high-definition television) [HAN 12], which itself relies on MPEG-2 technology, but is partly incompatible with DVB.

8.2.3.1. *DVB-S*

DVB is an initiative that answers the problem of analog television: normalization. In fact, analog television standards (PAL, SECAM,

D2-MAC, etc.) are not uniform and are sometimes incompatible [RAC 83]. The success of DVB cannot, therefore, be complete without a fairly strong impetus from European bodies to prompt the entire chain of actors, from design to production, to enter the consortium.

Currently, the DVB norm is well-installed across the entire world, particularly its satellite version. Competition remains high for media, such as cable and terrestrial waves [GIR 00].

Although DVB owes its success largely to that of MPEG-2 compression and the European initiative from which it originates, its adaptability is a fundamental element of its current and future success. In fact, the DVB system was designed to be broad and intelligent, adaptable to different methods of transmission to the extent that it imposes no restrictions on the material used. DVB was created to be able to transmit video like audio, as well as EPG (electronic program guides) and teletexts, and even has the means to transmit data in packet mode (datagram).

This open side of DVB is one of the elements that currently interest the different European and Asian projects, and this is what led us to study this standard.

8.2.3.2. *DVB standards*

DVB is a norm that in the first place governs cable transmission of digital TV, terrestrial waves or satellite. DVB standards are numerous and can be divided into three categories:

– Transmission standards that affect implementation of ES (elementary streams) transport MPEG-2 on different supports. The three main standards in this domain are DVB-S (satellite), DVB-C (cable) and DVB-T (terrestrial).

On the other hand, there is DVB-X (or DVB-M), paired with DVB-T; it affects commercial offers for mobiles (television on cell phones, video on demand) [MIA 15].

– Interaction standards that occupy return paths and connections (DVBDATA) contain all the standards DVB-RC (DVB – return channel).

Among them, DVB-RCS normalizes the return channel via satellite. We also find support for applications such as teletext (DVB-TXT) or subtitles (DVB-SUB) and finally a basic standard, DVB-SI (DVB – system information), which governs signaling for different programs present in the DVB flow, using signaling tables.

– Interfacing and encoding standards especially with DVB-CI (DVB – common interface) using conditional access (DVB conditional access) currently encrypted with DVB-CSA (DVB – common scrambling algorithm).

8.2.3.3. *MPEG-2*

The MPEG-2 compression norm forms the basis for norm DVB. Standard MPEG-2 also covers particular multiplexing methods relying on cutting down elementary flows (ES) into PES (packetized elementary stream) [GAR 11].

MPEG-2 offers two different methods for multiplexing PES depending on the application, forming two distinct types of flow.

The program stream (PS) is formed from a single program covering several PES (video, audio and private data). Formed of relatively long packets (2,048 bytes, for example), PS is used for transmission on a channel with a low error rate (quasi error free).

On the contrary, the second type of flow, TS (transport stream), can mix different programs. Moreover, it is the flow used to transport multimedia data on different supports that involve a non-negligible risk of error (cable, terrestrial waves, satellites, etc.). TS flow therefore uses small packets (fixed at 188 bytes) for the efficiency of corrector codes (Figure 8.1)

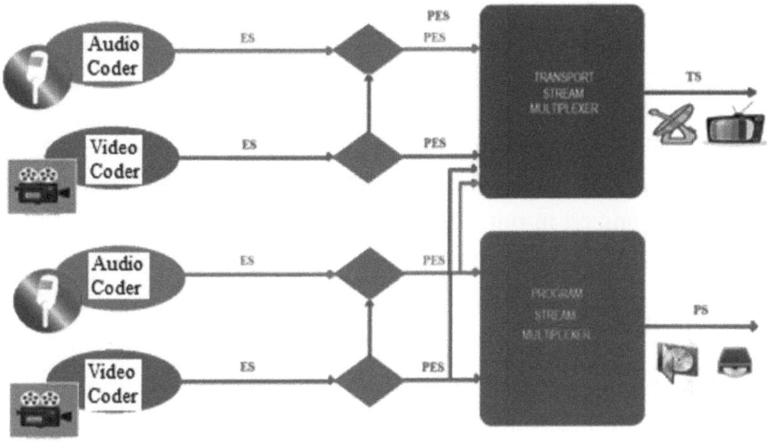

Figure 8.1. *Two types of multiplexing for MPEG-2*

8.2.4. *Satellite networks*

Satellite communication is the only means of providing a high bandwidth exactly where and when it is needed, independently of the geography and local infrastructure.

DVB-S [EVA 05] is a European standard designed for diffusing images and digital sounds, and also any other type of digital information via geostationary satellite. It relies on standard MPEG-2 TS [VUK 10] for compressing and encoding information, signal modulation, multiplexing audio and video signals and data synchronization.

DVB-S defines the method of multimedia traffic transmission by a one directional link via satellite to terminals equipped with a network interface DVB-S. Data other than video flows can be supported. It uses QPSK modulation and offers a flow of 40 Mb/s.

8.2.4.1. *DVB-S protocol stacks*

The DVB-S protocol stack relies mainly on norm MPEG-2. Figure 8.2 suggests an overview of this protocol stack for the user on the one hand, and for monitoring/management on the other [EVA 05]. These two stacks show the difference in processing between the flows MPEG-2 and signaling the flows. In fact, MPEG-2 flows use data streaming, which encapsulates the PES in MPEG-2 TS packets with a payload of 184 bytes (to which 4 header bytes must be added). Although signaling tables follow another process, MPEG-2 and DVB-S tables, organized in norm DVB-SI, are multiplexed in MPE (multi-protocol encapsulation) sections, which are themselves encapsulated in MPEG-2 TS packets and multiplexed with packets MPEG-2 TS containing MPEG-2 flows.

Data streaming for users and MPE for monitoring differ both from the method of inserting MPEG-2 TS packets, and on bit-stuffing management, as MPE can authorize encapsulation of two different sections in a single MPEG-2 TS packet; whereas data-streaming always uses bit-stuffing to complete an MPEG-2 TS packet.

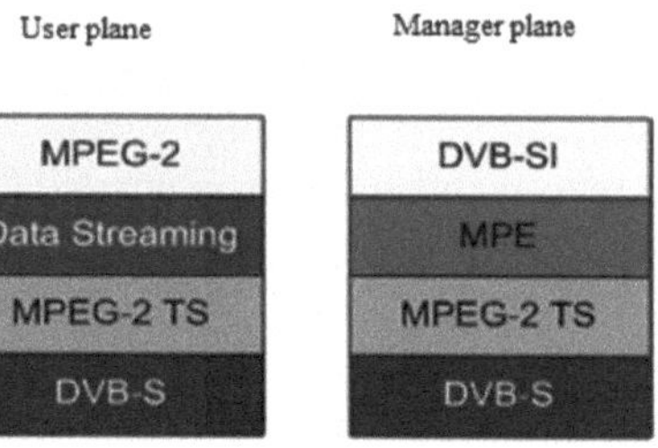

Figure 8.2. *Protocol stacks for the DVB-S as a support for MPEG-2 flows*

For IP traffic, the standard suggests encapsulating data packets in MPE, and then fragmenting them using MPE in packets MPEG-2 TS ready to be sent by the DVB layer.

8.2.4.2. *MPEG-2 TS encapsulation*

As described previously, MPEG2-TS packets have a fixed size, 188 bytes (payload of 184 bytes, 4-byte header).

However, the packet's header can be increased because of optional fields, at the same time decreasing the packet's payload. Figure 8.4 details the MPEG-2 TS packet and its standard header with 4 bytes [VUK 10].

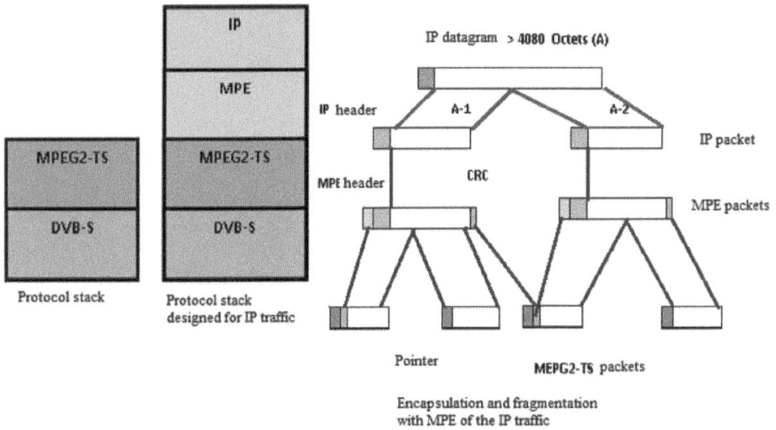

Figure 8.3. *DVB-S protocol stacks, fragmentation and encapsulation mechanisms*

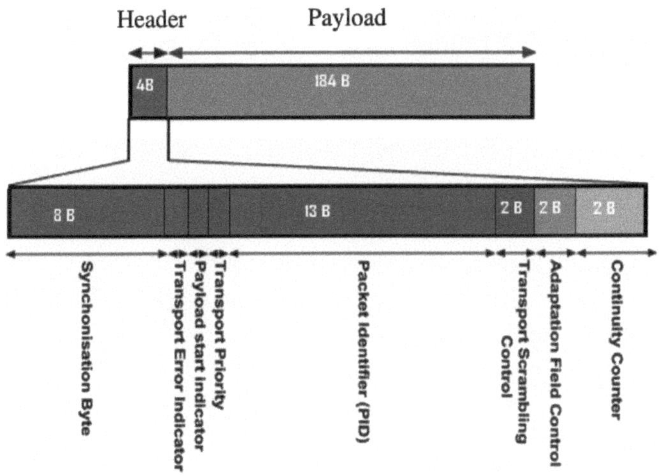

Figure 8.4. *Detail of an MPEG-2 TS packet and its header*

The fields of the MPEG-2 TS header are, in order of appearance, as follows:

– Byte synchronization: byte synchronization that takes the value 0100 0111 (0×47) and thus makes it possible to detect the start of a new packet.

– Transport_error_indicator: transport error indicator, 1 bit, indicates the presence of at least one incorrect bit that cannot be recuperated in the MPEG-2 TS packet.

– Payload_unit_start_indicator (PUSI): the PUSI (1 bit) shows if the start of a PES is present in this packet.

– Transport_priority: this is a 1-bit flag. When it is activated, the packet has greater priority than any other packet in the same PID.

– PID: formed of 13 bits, it is the channel's logic, that is, a stream number. Some PIDs are reserved (see section 2.3).

– Transport_scrambling_control: 2-bit field indicating the encryption type used for the TS payload.

– Adaptation_feld_control: this (2-bit) field makes it possible to know if an optional header called an adaption field follows the standard header:

- 00: reserved for future users;

- 01: no adaption field, only data;

- 10: only one adaption field, no data;

- 11: adaption field data monitoring.

– Continuity_counter: the continuity counter is formed of 4 bits; this field is incremented from one to each new MPEG-TS of the same PID. When it reaches the maximum, it returns to 0.

A PES flow is encapsulated using data streaming as illustrated in Figure 8.5. In data streaming, the PUSI is not used and the MPEG-2 TS packet is filled with stuffing. Thus, in the same packet there can only be one element of a single PES. The packets' PID will be the same for a given PES as well as the same for the same PES flow, that is, the original ES.

The stuffing used in DVB is considered as an adaption field. The control adaption field is therefore set at 11, and an adaption field is inserted, indicating the field size, followed by bytes of stuffing, thus making it possible to find the start of the data again easily (Figure 8.5).

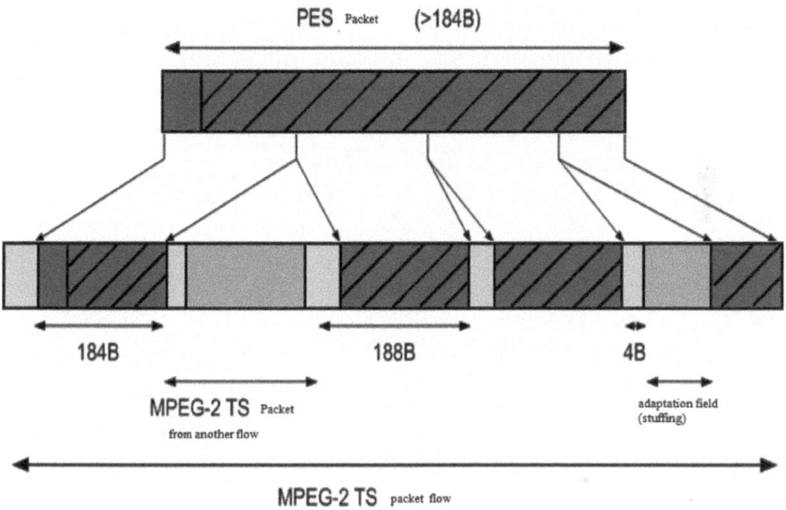

Figure 8.5. *Encapsulating a PES flow using data streaming*

The DVB-S system is formed of three distinct segments:

1) Segment operator formed of a terrestrial core network that transmits different flows to a gateway that is the central station responsible for sending recent flows received by the core network to the satellite, and also formed of the NCC (network control center) and the NMMC (network and mission management center), which are the system's management bodies (they work on access to the system as well as signaling) and can be integrated into the gateway.

2) The spatial segment is formed of one or more satellites responsible for diffusing the traffic received by the gateway to users.

3) The user segment is formed of reception terminals (very small aperture terminals [VSATs]) situated in the area collecting flow diffused by satellite and which sense the signals and process them to transform them into television programs for the user.

In some cases, access to the satellite system is provided by several gateways. These often rely on a frequency division in the bandwidth, or indeed a time division for on-board intelligence systems. The gateways transmit with a constant, permanent flow and the stuffing is therefore available for shared use when needed. DVB-S transmitters use a refined technology and are therefore high-cost, whereas DVB-S receivers cost little, and are the product of proven technology.

8.2.5. *Transmission chain*

Once the MPEG-2 TS flow is created, this is processed to be transmitted to the satellite by the DVB-S layer. The MPEG-2 TS flows are processed by the following encoding chain (Figure 8.6), which retraces the main stages of physical coding [VUK 10]:

– Stuffing: Also called energy dispersion, this makes it possible to avoid too long a succession from 1 or 0, which would result in the creation of a high-energy ray. The signal is multiplied by the output from a pseudo-random generator.

– Reed–Solomon coding: This coding adds a code of 16 headers to the end of the packet MPEG-2, making it possible to correct at most eight defective headers.

– Interlacing: This mix of data makes it possible to avoid consecutive sequences, which are hardest to correct, by interlacing the headers in several packets.

– Convolutional code: A lattice code, it adds redundancy to the signal at a rate of 2 bits for 1 (Viterbi decoding).

– Punching: This makes it possible to improve the convolutional code's performance by eliminating some bits (it is called forward error code for this performance).

– QPSK modulation: This makes it possible to modulate the signal without carrying out amplitude modulation as the signal is very noisy, due to the transponders' saturation work.

The flow symbol in the transmission chain is 27.5×10^3 symbols per second for a 35 MHz transponder; hence, there is a useful flow of 38 Mb/s for an FEC of 27,500.

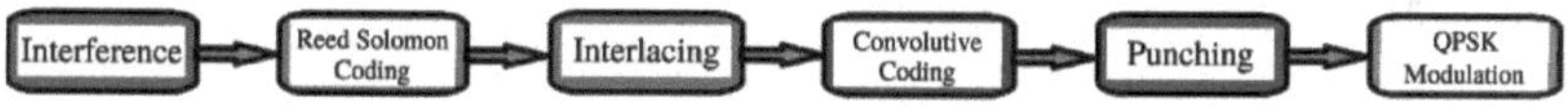

Figure 8.6. *Sketch of the DVB-S's coding chain*

8.2.5.1. *Service tables*

Signaling tables are needed to manage transmission and reception of different flows and programs carried on the satellite. These tables are called service tables and are defined in several standards:

– ISO MPEG-2 [AHL 06] defines PSI (program-specific information) tables, which describe the different elements present in an MPEG-2 TS flow;

– DVB-SI [AWA 07] suggests tables specific to the DVB system. The SI tables provide additional information to the PSI tables, indicating at the terminals the different services offered by the system;

– Other tables are envisaged by the DVB norms. Interactive norms such as DVB-RCS add signaling to manage the return channel (capacity and access).

8.2.5.2. *DVB-S2*

DVB-S2 [MIC 12] is the most recent spatial satellite communication norm in the DVB family. It aims to improve the satellite's video transmission capacity. To do this, it suggests using BCH (Bose—Chaudhuri–Hocquenghem) codes and LDPC (low-density parity-check) codes for the channel coding. It also defines four modulation modes: QPSK, 8APSK, 16APSK and 32APSK.

The norm can be extended to cover several types of traffic such as IP, unlike the former DVB-S norm that concentrates on MPEG-2 traffic. This norm's only drawback is the same problem that new technology always encounters: the installation of new components, given that DVB-S2 is not compatible with the first.

Finally, DVB-S2 suggests a new video coding norm: H264/AVC [RE 13] in place of MPEG-2 TS, in the case of DVB-S for a better QoS. Norm H264/AVC aims to improve the gain in coding efficiency with a ratio of at least 2 compared to existing standards (MPEG-2, H.263 and MPEG-4 Part 2), and aims to create a simple interface to be able to adapt the Codec to different transmission protocols (switching packets and circuits).

8.2.6. *Satellite system (DVB-RCS)*

8.2.6.1. *Introduction*

The DVB-S and DVB-S2 norms [AIJ 13] do not specify any recommendation for installing a return channel by satellite. For interactive services, a link is established, either by land links via telephone lines, or by satellite. To do this, DVB-RCS [LI 11] defines a return channel from the users to the satellite. Paired with DVB-S they together offer a two-directional (2D) satellite system.

DVB-RCS is therefore considered as an addition to DVB-S. In fact, the flows leaving the DVB-S have different information to govern the global system in terms of access, signaling and synchronizing terminals. The DVB-RCS system defines a 2D broadband system by satellite for VSATs such that the flow leaving the terminals follows DVB-S (or DVB-S2) and the return channel follows the technical specifications for the air interface. DVB-RCS has been specially created for an interactive channel. It reduces the cost of the link, offers users a return satellite link without uselessly monopolizing resources, which is not the case with DVB-S, which transmits a constant data flow and triggers stuffing in the absence of any data to transmit. DVB-RCS is presented as a contribution to DVB-S, and it is one of the reasons why the two norms are often confused, as they work in direct correlation. It is therefore natural to introduce it here by incorporating it with a DVB-S architecture, as the norm specifies.

8.2.6.2. *DVB-RCS architecture*

According to the specifications of this norm, a DVB-S system is able to interact with a DVB-RCS system.

This means that the system can be supported by two different satellites (Figure 8.7), or a single satellite that includes both technologies. The DVB-RCS system is integrally dependent on the flow leaving the DVB-S, which contains the different information to govern the system (access, signaling and terminal synchronization).

The method envisaged by the norm thus makes it possible to remain in the current DVB-S framework while still bringing an interactive margin to different applications. The two advantages of technologies are thus combined in a single system, which however has this nuance: even if the technological cost of a DVB-RCS transponder is less than that of a DVB-S transmitter, especially with VSATs, the material costs of installing and locating the bandwidth remain high.

DVB-RCS uses a star-shaped network architecture (Figure 8.7) as all traffic created by the user terminals (return channel satellite terminal [RCST]) transits through the NCC gateway, which acts as a gateway between the RCSTs and Internet access providers [93]. The RCSTs can transmit and receive data. The control center manages access to the system and much of the signaling. For the next generations of satellites, the system's management procedures could be installed onboard the satellite.

On the other hand, to provide Internet access for users located in an area not covered by an infrastructure network, RCSTs can be considered a relay for IP traffic.

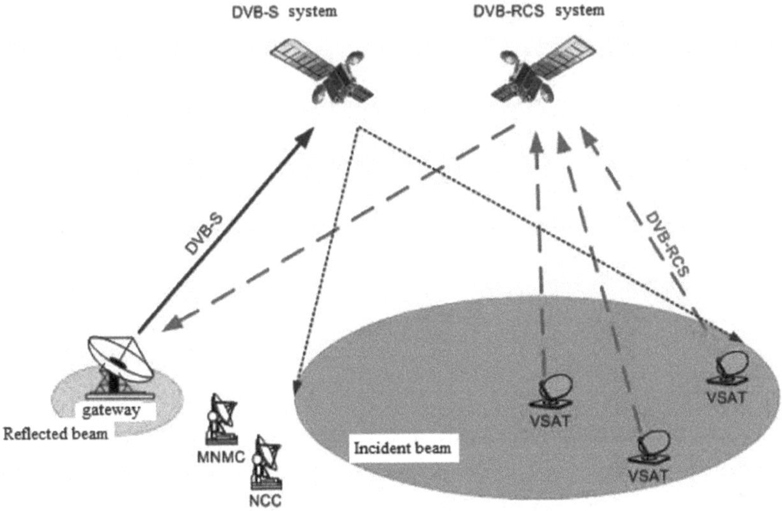

Figure 8.7. *DVB-RCS architecture [POL 03]*

8.2.7. *Protocol stack*

The DVB-RCS norm has two possible protocol stacks (Figure 8.8) [OLI 10]. These stacks are different at layer 2 in that there are two methods of encapsulating the data to be transmitted: a method that uses ATM and another based on MPEG-2 TS packets. However, encapsulation by the AAL-5 (ATM adaptation layer) is the obligatory technique that each RCST should be able to use. The second method of encapsulation, defined by the MPE layer, is optional [LI 11].

DVB-RCS is based on the transmission of bursts using a specific breakdown into time and frequency and dependent on the type of data to be transmitted. DVB-RCS suggests two encapsulation methods: a method using ATM cells and another relying on MPEG-2 TS packets. As Figure 8.8 shows, the protocol stacks used to encapsulate standard data (here an IP datagram) differ from the chosen methods.

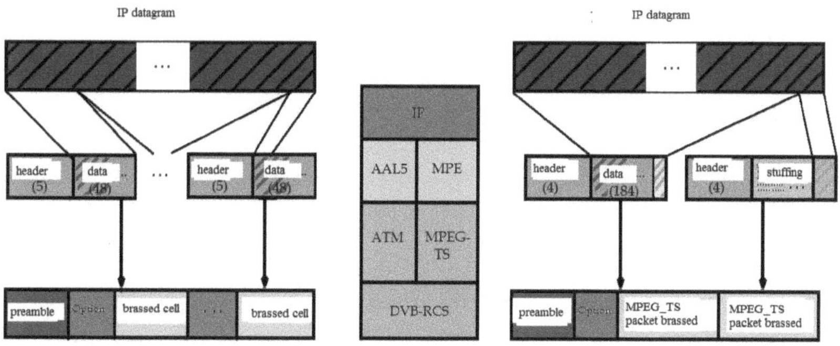

Figure 8.8. *Two encapsulations suggested for DVB-RCS combined*

The difference between these two piles is found in layer 2, as can be seen in the following:

– Encapsulation in ATM cells is managed by AAL-5, which is adapted for encapsulating datagrams. This technique is described by the norm as obligatorily supported by any system using DVB-RCS technology.

– Encapsulation in packet MPEG-2 TS uses the sections method used to encapsulate SI and PSI tables: the MPE protocol. This method is optional depending on the specifications of the norm.

In both these cases, the physical layer remains the same: the DVB-RCS. Once encapsulated in cells or packets, the data are grouped in a burst (one, two or four cells generally) are then mixed to disperse energy. A preamble is then applied to the packet depending on the different encodings. Finally, we will note the presence of an optional field, which is used particularly in the SAC (satellite access control) method.

8.2.8. *Physical layer*

The data (medium access control level data or traffic) are formatted in slots (bursts) for transmission. We distinguish four types of slot:

– the TRF (TRaFfic) traffic slot: this slot contains data, ATM or MPEG-2, from the terminals;

– the CSC (common signaling channel) slot is responsible for identifying terminals during the log-on stage;

– the ACQ (ACQuisition) slot enables coarse synchronization of the terminals;

– the SYNC (SYNChronization) slot enables finer synchronization of the terminals and transport information such as connection requests, capacity requests, etc. As for the DVB-RCS transmission chain, this norm suggests using turbo-codes as an option [OLI 10]. In fact, as Figure 8.9 illustrates, according to CRC (cyclic redundancy check) encoding, we can either pursue processing identical to the DVB-S's classic transmission chain, or replace the Reed–Solomon encoding chain, interlacing and convolutive encoding (also called concatenated coding) using a turbo-code [TAL 11].

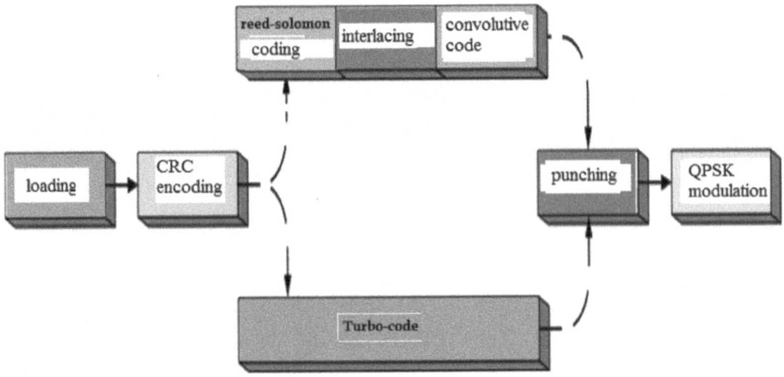

Figure 8.9. *Overview of the DVB-RCS encoding chain*

It should however be noted that the general trend is toward the use of turbo-codes in DVB-RCS. These codes, which are more complex than convolutional lattice codes, are also much more effective [ZHU 11].

8.2.9. *Access method*

8.2.9.1. *Sharing resources*

Unlike the DVB-S norm, where there are a very limited number of user gateways accessing the result, in the DVB-RCS norm, a considerable number of terminals share the network's resources [BER 96]. The DVB-RCS norm uses TDMA (time division multiple access) mode only for broadcast transmission in the case of a regenerative payload; TDMA poses a problem on the return. In fact, for the broadcast channel, there is more often a single gateway that is content to cut the allotted band into different user groups, whereas in the context of the return, a large number of terminals wish to share the same channel. Also, the TDMA is not sufficient to avoid collisions between the different bursts from the active terminals.

Because of this, the return DVB-RCS is based on the MF-TDMA mode (multi-frequency TDMA), which uses a division into the allotted band using super frames. A super frame is allotted to each group of terminals and this is itself cut into frames. Each frame is formed of a set of time slots, each used to transmit a precise burst. Figure 8.10 shows this division.

There are two modes of sharing MF-TDMA: static mode and dynamic mode.

In the static mode, we use a frame-by-frame MF-TDMA configuration. This technique greatly reduces the waiting time for a media transmission and for the messages to be sent. However, in the dynamic mode, the MF-TDMA structure makes it possible to vary the time slots, thus enabling greater flexibility by giving the NCC the chance to allot more or fewer large time slots, depending on the speed of the traffic. However, this method is complex and requires more calculations on the part of the NCC (it should transmit information from each time slot), which risks degrading the access time to the support.

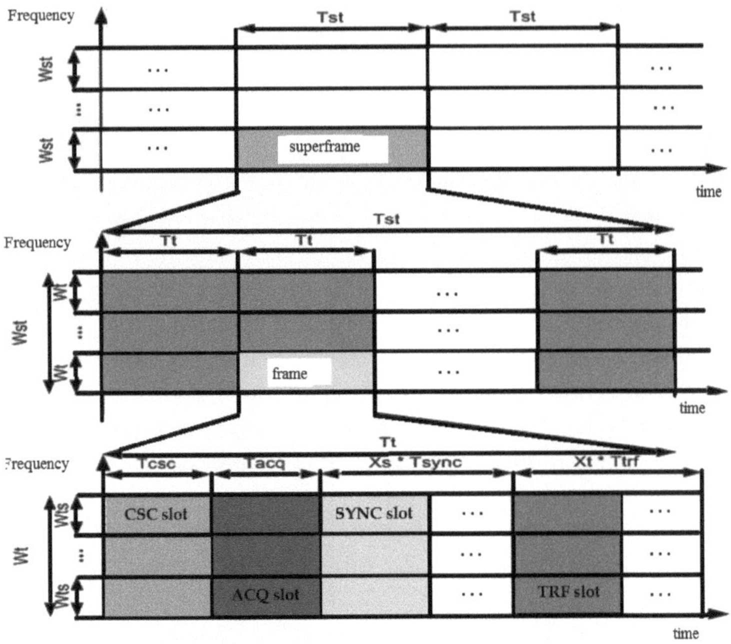

Figure 8.10. *Segmentation of the allotted band on the return, in super frames, frames and time slots*

8.2.9.2. *Resource allocation*

Resource allocation happens at two levels: at the connection via the CAC (connection admission control) function and the MAC by the DAMA (demand assignment multiple access) function. The CAC function makes it possible to access support for the duration of the connection by keeping some of the resources. The DAMA function makes it possible to ensure allocation of time slots at the terminal to check the QoS parameters (throughput, jitter, transfer time, etc.). These two functions are thus used on two different time scales.

We emphasize that there can be several types of resource allocation methods. Access to support can be fixed for the entire duration of a communication, allowing a terminal or a group of users a set part of the bandwidth. The bandwidth for a communication can also be allocated dynamically, thus making it possible to adapt to the traffic, and optimize use of the satellite capacity. These methods are however more complex to implement and introduce access time for support that is variable and complex to model.

8.2.10. *Deploying IP by satellite IPoS*

IPoS (IP over Satellite) [LI 11] is a standard that was developed by group TR 34.1.7 of the Telecommunications Industry Association (TIA). It relies on the HNS'DIRECWAY system, specifying techniques for transporting IP packets between a central hub terminal and distant terminals using geostationary satellite.

IPoS operates in the Ku band, but it can extend to cover other frequency bands. IPoS specifies an access point to an independent satellite service (SI-SAP) that creates a well-defined interface between the functions dependent on the satellite and the application layers, thus enabling activation of an open service platform. The protocol stack suggested for the IPoS as well as the description of the protocol layers are presented in Figure 8.11 [BER 96].

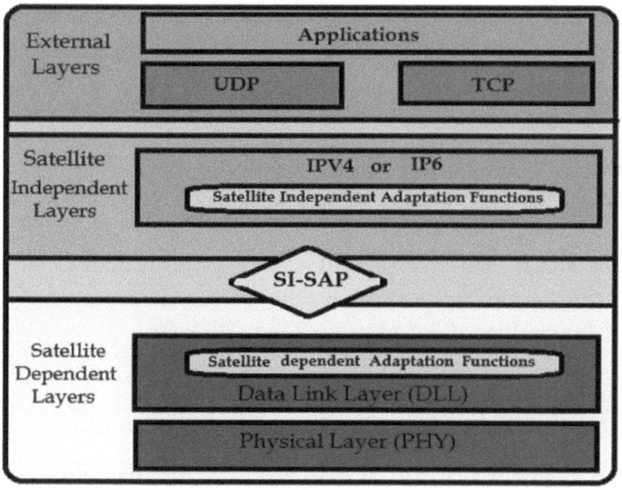

Figure 8.11. *IPoS protocol layer*

The architecture of IPoS is divided into three segments:

– hub segment: it supports Internet or Intranet access for a large number of distance terminals via satellite. It is formed of a large terrestrial hub and its equipment for transmitting data flows to the satellite;

– spatial segment: this is a geostationary satellite that enables transmission in both directions between the hub and the terminals;

– distance terminal segment: it contains user terminals, usually formed of two parts: an outdoor unit, ODU, and an indoor unit, IDU.

The architecture of the IPoS is also formed of three interfaces:

– the user interface: it is situated between the terminal and the user equipment; this is generally the LAN interface;

– the IPoS interface: it is between the satellite and the hub, and between the satellite and the terminals. It specifies transmission in both directions: from the hub to the terminals, and from the terminals to the hub;

– the hub's terrestrial interface: it is between the hub and the outdoor units that switch packets.

8.2.11. *Methods of encapsulating IP packets*

Encapsulation of IP packets is needed for these packets to be able to be transported by the DVB-S in MPEG-2 segments. Among existing encapsulation methods, there are two standard methods available, which are MPE and ULE (unidirectional lightweight encapsulation), and a specific method, the "IP-optimized scheme" suggested for multicast communications [BER 96].

In this chapter, we focus on the ULE method.

8.2.12. *ULE encapsulation and header extensions*

ULE encapsulation creates the SNDU data frame by adding a "header" of 10 bytes and a tail of 4 bytes at the PDU network layer. The ULE frame is shown in Figure 8.12.

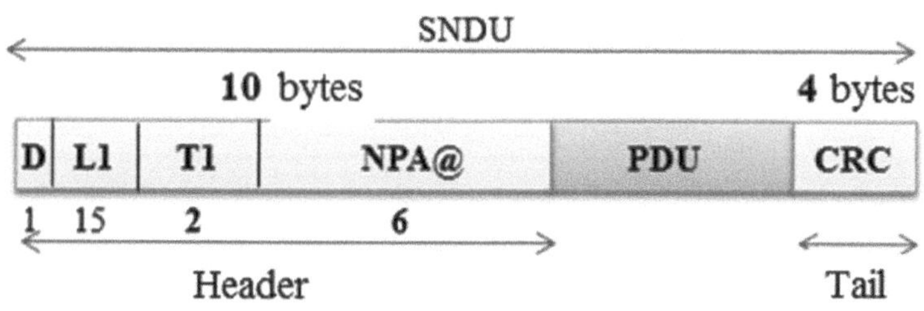

Figure 8.12. *ULE data frame*

The header is formed of four fields, which are as follows:

– the *D* field: formed of a single bit (0 or 1) indicating the presence or absence of the optional field NPA@ (Network Point of Attachment address);

– the length field *L*1: formed of 15 bits, indicating the length of the SNDU frame in bytes, from the first byte that follows the *T*1 field until the end of the last byte in the tail;

– the type *T*1 field: formed of 16 bits, indicating the type of PDU carried by the SNDU or the type of extension on the tail if it exists. In the first case, T1 takes a value greater than 0×0600 (e.g. $T1 = 0 \times 0800$ if the PDU is an IP packet), and in the second case, it takes a value lower than 0×0600;

– the optional field NPA@: formed of 6 bytes, contains the destination address of the SNDU frame. Usually, it is a MAC address similar to that used by the Ethernet protocol.

The tail is a CRC-32 code applied across the SNDU.

If we wish to add additional services, such as security, we need to add additional information. ULE encapsulation has a very supple mechanism that makes it possible extend the original header as shown in Figure 8.13.

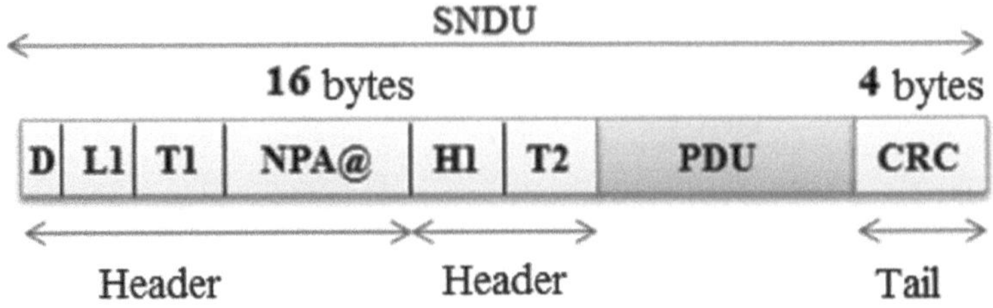

Figure 8.13. *ULE encapsulation with an extension on its header*

8.2.13. *Interconnection of WiMAX mobile and satellite DVB-S networks: RCS*

In this section, we focus on a model for interconnection between the satellite network formed of two combined systems: the DVB-S/RCS and mobile WiMAX. As we have mentioned earlier in Chapter 3, Macromobility of level +3 is ensured by the SCTP.

This protocol is based on the concept of multihoming for managing mobility. With its method of creating several connections at the same time (following the IP addresses as well as available network interfaces), multihoming enables transparent passage in an HO zone. It only remains to us to define the transport protocol that suggests the most optimal QoS for interconnection of a satellite WiMAX mobile. To do this, we will base this model on three protocols: SCTP, MSCTP as well as CSCTP.

These protocols are installed on an MS that moves with two means and fast speeds, with the aim of demonstrating the effect of speed on the QoS obtained by such a proposition.

8.2.14. *Interconnection model and mobility scenarios*

WiMAX mobile and satellite interconnection with norm DVB-S/RCS is formed of a WiMAX cell with a coverage radius of 3 km as well as a zone covered by a geostationary satellite with a radius of 50 km.

The intersection of these two cells forms an HO zone, by passing through this zone the MS should be able to connect with both networks at the same time. This involves using both the WiMAX mobile interfaces explained in Figure 8.14 and another satellite. The latter's main purpose will be ULE /MPEG-2 TS encapsulation or de-encapsulation of the numerical IP packets that it receives or transmits, as well as access routing for the upstream traffic (attribution of priority levels to IP flows on return and appropriate processing). The satellite responsible for diffusing the traffic at the MS in the coverage zone communicates directly with an NCC gateway on the ground and in turn is connected to a router that ensures Internet access.

As for the WiMAX network, the BS is connected to an ASN-GW and the latter is linked to the CSN, which provides direct access to an Internet network.

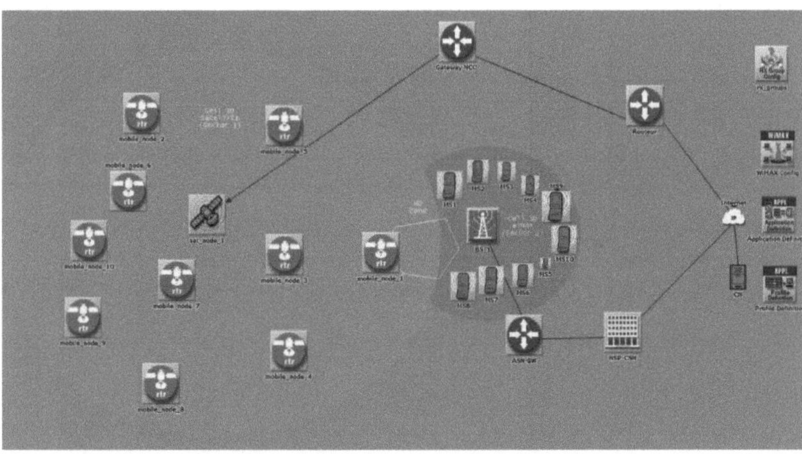

Figure 8.14. *Model for interconnecting WiMAX mobile satellite*

8.2.15. *Simulation scenarios*

The aim of this study is to define the version of the SCTP that ensures a better QoS during passage of the MS from one satellite network to another WiMAX mobile with complete transparency and vice versa. To do this, the MS should make several journeys from the satellite network to WiMAX and back throughout the entire simulation. Because the network resources are limited, to get closer to reality we install several other mobile stations in the satellite and WiMAX coverage zones. These mobile stations start to communicate with their respective networks 20 minutes from the start of the simulation. We also program a traffic-generating application between the MS and CN to transit a maximum flow of 12,000 bit/sec.

With the aim of validating our work's qualitative contribution, we carry out a comparative study between three different scenarios.

– *First scenario*: The MS and its correspondent (CN) use SCTP as a transport protocol.

– *Second scenario*: The MS and CN use MSCTP as a transport protocol.

– *Third scenario*: The MS and CN use CSCTP as a transport protocol.

To show the impact of increasing the speed on the MS's performances, it moves at two speeds: 50 and 100 km/h.

8.2.15.1. *Implementation with simulator*

In the first and second chapters, we activate the SCTP and MSCTP at transport layer 4, respectively. We return to the same modifications carried out in the previous chapters for installing SCTP and MSCTP in the simulator, for MS and CN.

For the third scenario in layer 4, we program the CSCTP. To do this, we add a new handoff state to the MSCTP's operation (see Figure 8.15), whose main function is to inform the CN when a handoff mode commences at the MN. This modification is made with the aim of quickly indicating the change of address to the CN. The CN considers both the known addresses (the old primary address and the new, added address) as the primary address; in this case, the CN sends duplicated packets on the two primary addresses to the MN. The association's old IP address is deleted when the MN decides that an address is inactive.

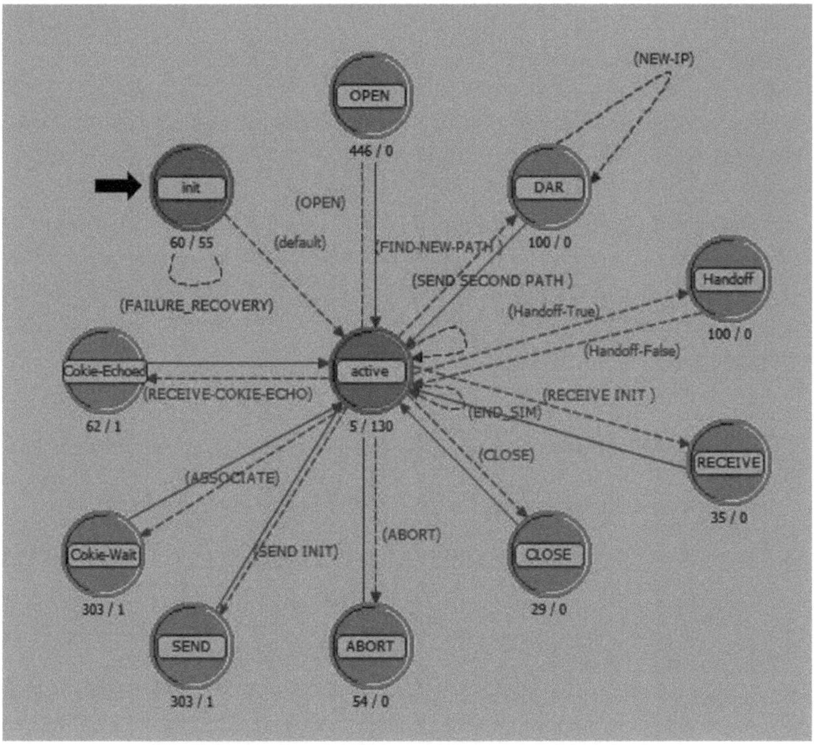

Figure 8.15. *Process model of the CSCTP*

Finally, to ensure correct running of the simulation, the MS's and CN's network interfaces should both install the same transport protocol for each scenario. Because the MS moves between the two WiMAX mobile and satellite networks, this means that activation of the transport protocol in the scenario occurs on the two communication interfaces. We take the example of satellite network interfaces used in the third scenario, based on the CSCTP (see Figure 8.16).

To enable the MS to receive and transmit packets supporting the SCTP/IP protocol in a DVB-S/RCS satellite type system, the data packets should be type MPEG-2 TS as we have already mentioned previously. Solving this difficulty means using an encapsulation mechanism at the link layer.

According to [STE 01], the ULE encapsulation mechanism is most appropriate for satellite communications, with SCTP/IP-type flows.

Encapsulating SCTP packets by ULE creates another MPEG-2 TS type packet (see Figure 8.15, packets using the ULE process) to be sent by the physical layer DVB-S from the NCC gateway to the satellite, and by the DVB-RCS physical layer from the MS's satellite interface (see Figure 8.16).

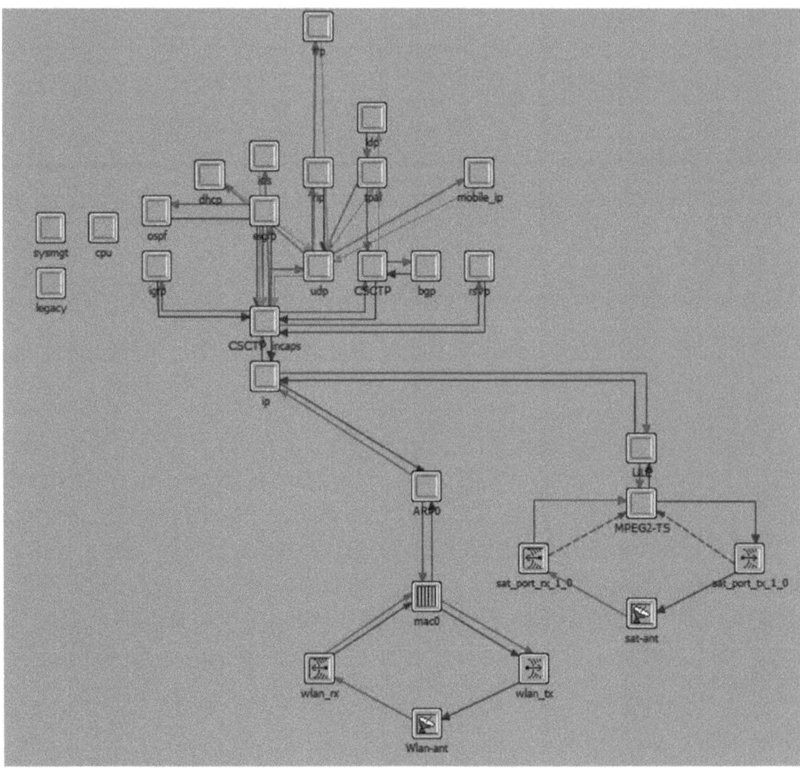

Figure 8.16. *Satellite interface using the CSCTP, scenario 3*

8.2.15.2. *Simulation parameters for the satellite network*

Sat-node parameters	
Orbit	Low-earth
Altitude (km)	1,000
Cycle time	60
Jammer band base frequency	30
Jammer bandwidth	10
Jammer transmitter power	100
Number of frequencies	60
NCC gateway	
Processing scheme	Central processing
Datagram switching rate (pkt/sec)	500,000
Datagram forwarding rate	150,000
Maximum static routes	4,000
Timeouts SCTP, MSCTP, CSCTP	86,400
Dynamic translations	86,400
Mobile node	
Wireless LAN MPEG-2 TS data rate (Mbps)	11
Transmit power (W)	0.005
Packet reception power threshold (dBm)	–95
Buffer size (bits)	256,000
Large packet processing	Drop
Long retry limit	7
Short retry limit	4

Table 8.1. *Simulation parameters for the satellite network*

For the WiMAX mobile network, we use the simulation parameters already mentioned (see Table 8.1).

8.2.15.3. *Performance criteria*

To compare the performances offered by the SCTP, MSCTP and CSCTP in vertical HO between the WiMAX mobile and satellite, we suggest the mean output received by the MS as performance criteria.

8.2.16. *Results*

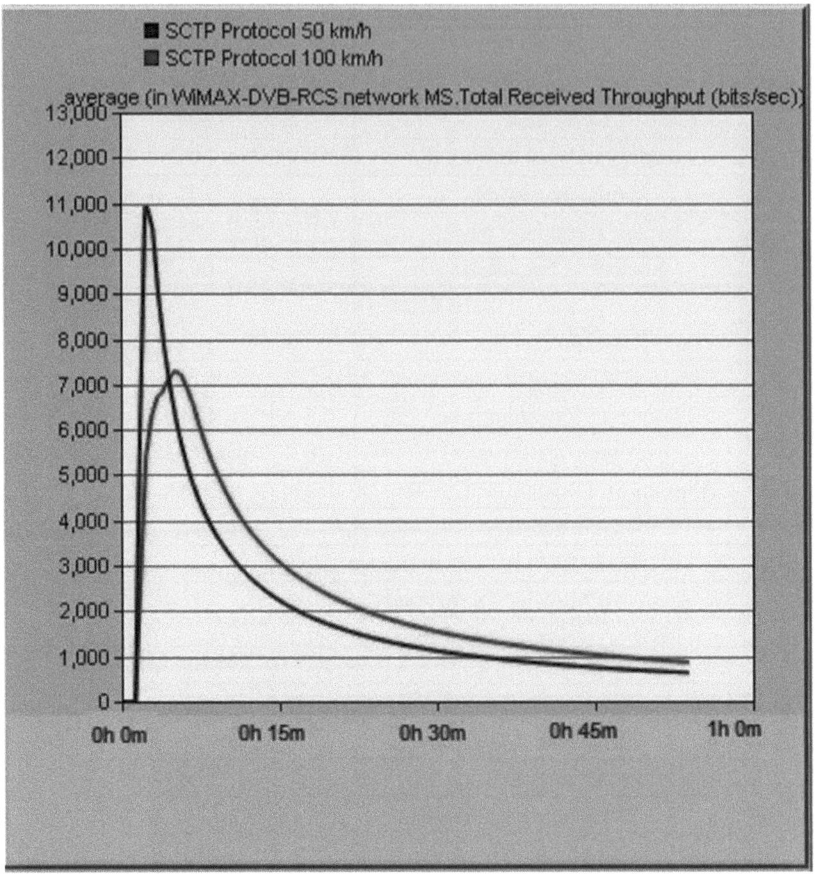

Figure 8.17. *The mean output for the SCTP for speeds of 50 and 100 km/h. For a color version of the figure, see www.iste.co.uk/benslama/networks.zip*

In Figure 8.17, we see the mean output received by the MS in bit/sec, using the SCTP (scenario 1). So we see that using this protocol creates a significant rise in the output, which reaches 11,000 bit/sec for a speed of 50 km/h and 7,000 bit/sec for 100 km/h at the start of the simulation. Then the output drops very rapidly to less than 900 bit/sec at the end of the simulation; as we mentioned earlier, the network traffic generator application transmits a flow of 12,000 bit/sec between the CN and the MS. Nevertheless, the version based on the SCTP reaches the maximum flow only at the start of the simulation, for a very short time.

Consequently, we henceforth concentrate on using the MSCTP as well as the CSCTP.

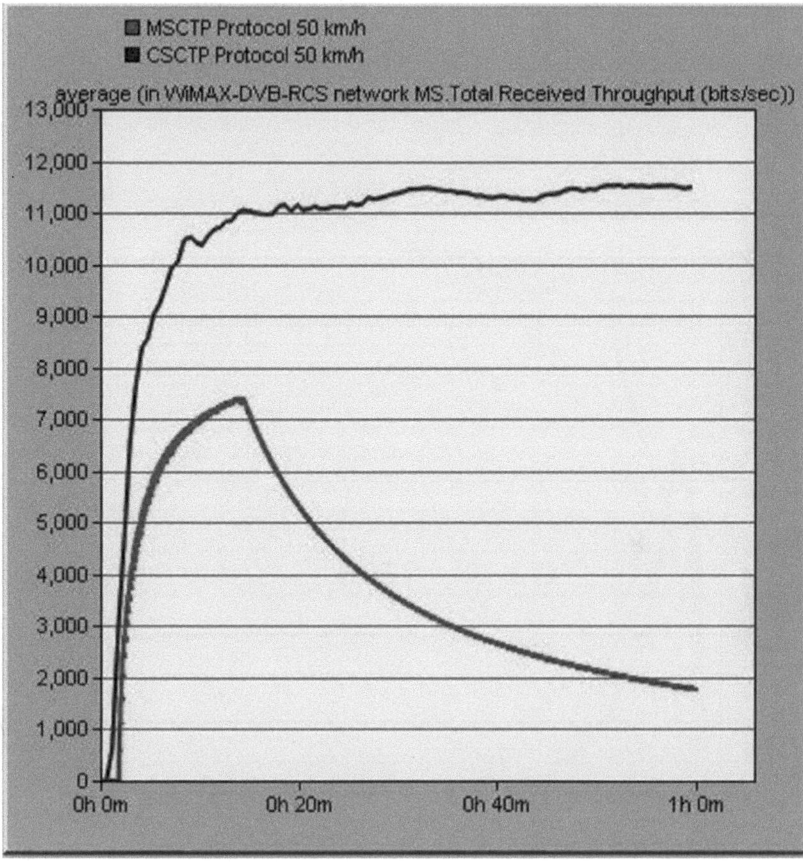

Figure 8.18. *The mean output for the MSCTP and CSCTP for a speed of 50 km/h. For a color version of the figure, see www.iste.co.uk/benslama/networks.zip*

Figure 8.18 shows a comparison between the mean output received by the MS compared to scenario 2 (using the MCTP) and scenario 3 (using CSCTP). This comparison is made using a speed of 50 km/h, and we see that the two scenarios show differences in output from the start of the simulation. This difference in output increases very quickly with time. However, after the 18th minute of simulation, the output using MSCTP starts to decrease to eventually drop to less than 2,000 bit/sec.

The output from the CSCTP, however, continues to increase to reach a maximum value of 11,563 bit/sec at the end of the simulation.

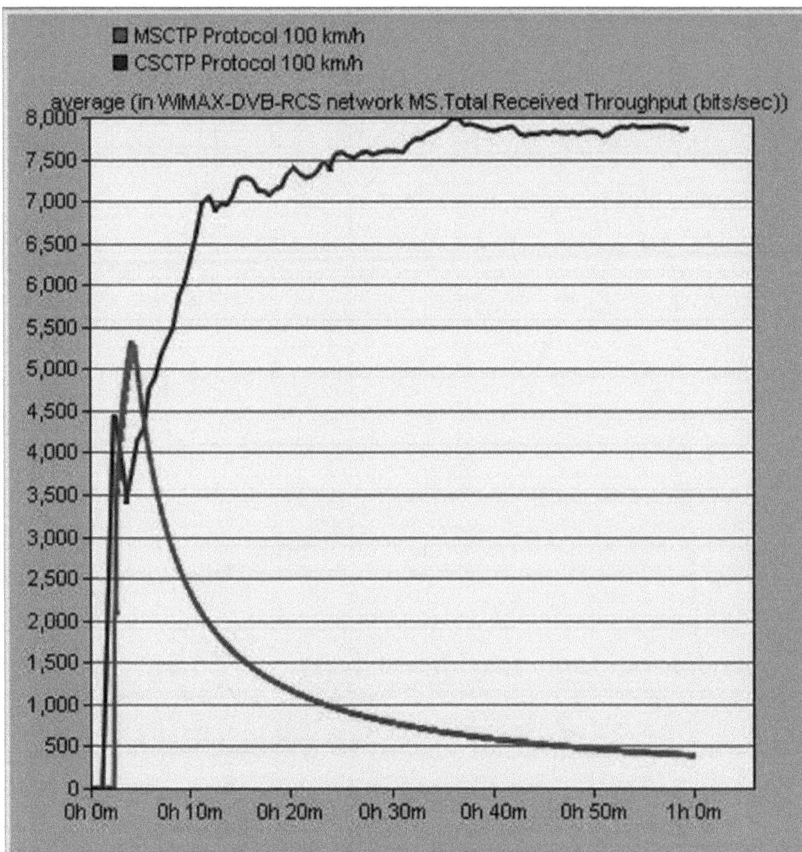

Figure 8.19. *Mean output from the MSCTP and CSCTP for a speed of 100 km/h. For a color version of the figure, see www.iste.co.uk/benslama/networks.zip*

Similarly, Figure 8.19 shows a comparison between the mean output received by the MS compared to scenario 2 (using MSCTP) and scenario 3 (using CSCTP). However, this comparison is made using a speed of 100 km/h, and we see that the two scenarios present differences in output from the start of the simulation. MSCTP reaches a peak output with a value of 5,398 bit/sec, to then drop very quickly to less than 500 bit/sec. However, CSCTP continues to increase to reach a maximum value of 8,000 bit/sec.

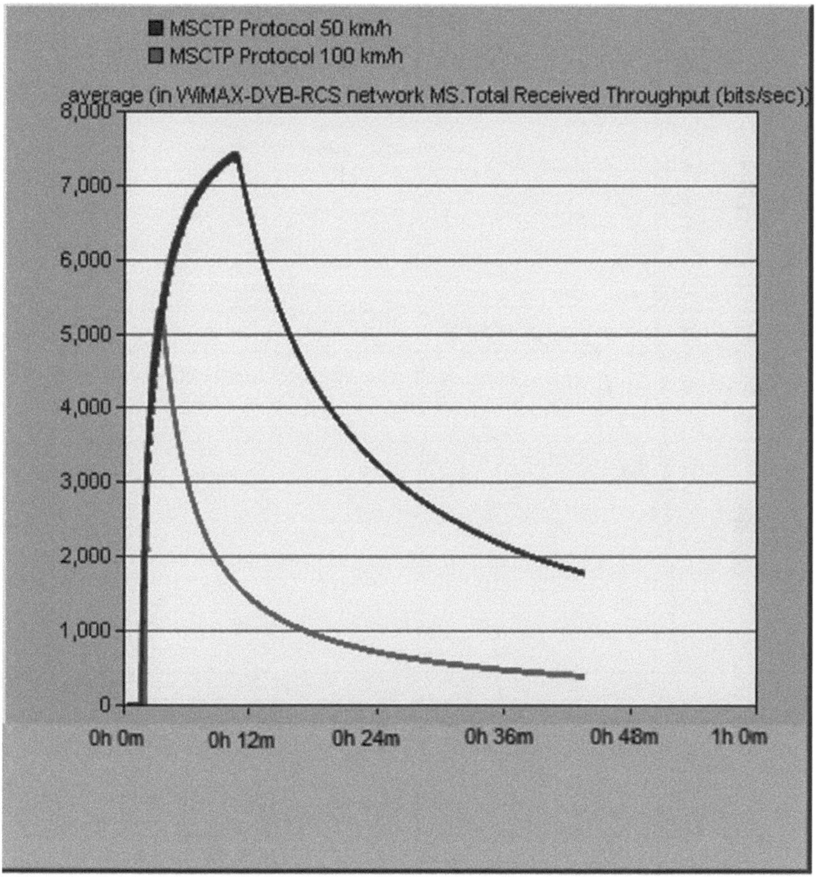

Figure 8.20. *Comparison of mean output from the MSCTP for a speed of 50 and 100 km/h. For a color version of the figure, see www.iste.co.uk/benslama/networks.zip*

Figure 8.20 shows a comparison between the mean output received by the MS compared to scenario 2, for speeds of 50 and 100 km/h. Even if the effects of the speed are visible from the start of the simulation, the two variants of this scenario show a relatively low output. This output reaches a maximum at the start of the simulation, to then drop to less than 2,000 bit/sec for a speed of 50 km/sec and 500 bit/sec for a speed of 100 km/sec at the end of the simulation.

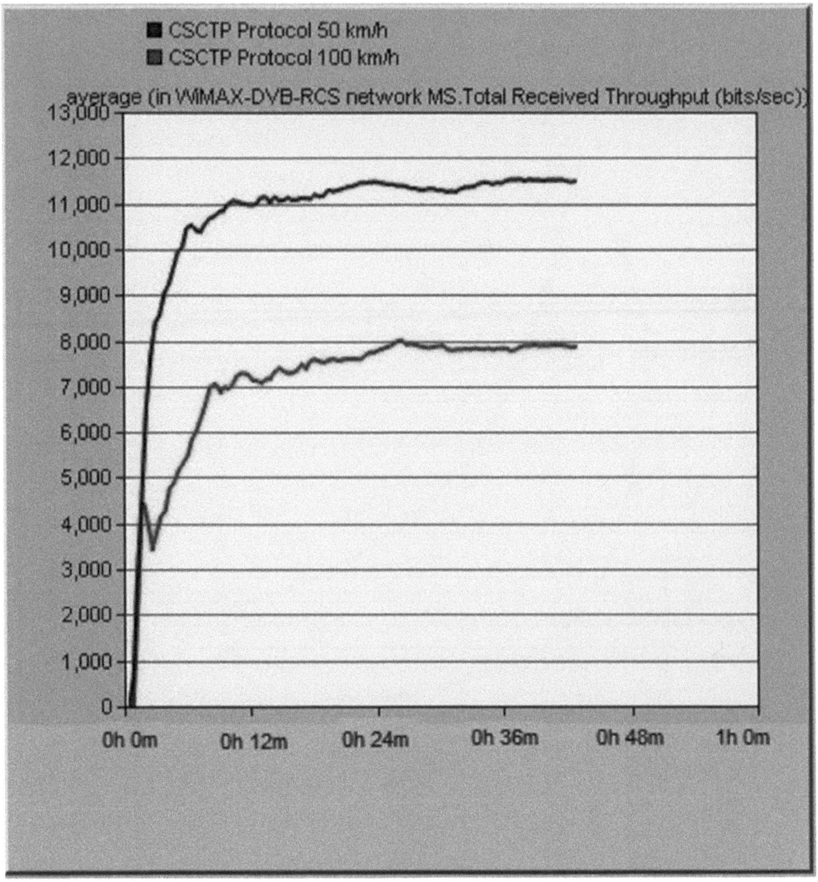

Figure 8.21. *Comparison of the mean output from the CSCTP for speeds of 50 and 100 km/h. For a color version of the figure, see www.iste.co.uk/benslama/networks.zip*

As the CSCTP presents a better output than SCTP and MSCTP, in Figure 8.21 we compare the output of an MS installed on the CSCTP and moving with a speed of 50 and 100 km/h. We see the effects of the speed from the start of the simulation, as the two variants of scenario 3 have the same value only for a few seconds. Then, the difference in output between the two speeds widens; for a speed of 50 km/h, the output registers a maximum value of 8,000 bit/sec, whereas with a speed of 100 km/h it reaches 11,563 bit/sec.

8.2.17. *Discussion of the results*

In a WiMAX mobile and satellite during a vertical HO networks, the transport protocol plays a fundamental role for creating transparency of passage during a vertical HO between the two networks. We have seen this by comparing the different scenarios shown in this chapter. As we have already noted, SCTP and MSCTP are not reliable protocols for a vertical HO between WiMAX mobile and DVB-RCS satellite, as the present reliable flows that are aggravated by the effect of increasing the speed, as well as the number of subscribers in the two networks' coverage zones.

As for CSCTP, this protocol has proven to be very resistant to the effects of a vertical HO between WiMAX and satellite networks as it achieves a satisfactory mean output. Nevertheless, increasing the speed of movement involves decreasing the mean output by nearly 33%.

We explain these results by the speed with which the CN is informed when a handoff mode is triggered by the MN using the bit H of chunks flags. This modification enables the CN to consider simultaneously both the paths defined by the MN (the old and the new) as primaries.

In this case, the CN sends duplicated packets on the two primary addresses to the MN. With the substantial delays satellite communications involve, for MSCTP the data packets are sent from the old IP address before the MN considers the new IP address as a primary address for the association underway.

In this context, using multihoming is not favorable to increasing the output, which makes the MSCTP's flows comparable to those of SCTP.

8.2.18. *Conclusion*

Because of the great difference in performance between the WiMAX mobile and satellite systems, interconnection between them has proven to be very complex. In this chapter, we have suggested an interconnection model between a spatial network based on the principle of diffusion by satellite, and formed of two combined systems: the DVB-S/RCS and a WiMAX mobile network. With the aim of achieving transparent passage between these two systems with the both constraints of speed and of the number of subscribers, we have based our study on the use of three transport protocols: SCTP, MSCTP and CSCTP.

The results of simulations obtained from our model show that the CSCTP is better adapted for a vertical HO between WiMAX mobile and satellite networks. Thus, use of the multihoming concept is more favorable for the output in the context of using the CSCTP than for the MSCTP, which suggests results comparable to those of the SCTP.

9

Compressed Sensing in Interconnections Covering WiMAX, UMTS and MANET Satellite Networks

9.1. Introduction

Each of the telecommunication technologies that have come about has been designed independently to support different types of data, coverage or mobility. These telecommunication systems have their advantages and disadvantages. No technology can simultaneously provide services with a high bandwidth, high mobility and a broad service zone for a large number of users.

To do this, an interconnection method makes it possible to use the advantages of all the technologies combined and thus reduce their drawbacks.

In this chapter, we focus on studying a model for interconnection covering the four networks studied previously, that is, DVB-S/RCS satellite, WiMax mobile, UMTS and MANET. We focus on creating a centralized infrastructure for managing mobility at level 3+ performance optimization for the MS.

9.2. Heterogeneous networks

The term "heterogeneous networks" comes from the integration of different telecommunication technologies with the aim of using their additional characteristics. Because of heterogeneous networks, users can transport the connection between several access points of different types, benefitting from the best customized services depending on their personal preferences. Moreover, heterogeneous networks represent an important solution for decreasing congestion on mobile networks by sharing traffic with other access technologies with higher flows. In this study, we focus only mobile and satellite heterogeneous networks.

Because of their fast and significant deployment, mobile networks will consist of a set of heterogeneous systems, managed by different operators and formed of distinct access networks [HAA 99]. In this context, mobile terminals will be network multi-interfaces, which enables them to move from one system to another transparently during communication. We therefore speak of global mobility and vertical handover.

All the user connections with this heterogeneous network should occur transparently: without interruption and without degradation in service when the user changes from one network to another. To do this, networks should be based on existing infrastructure, by interconnecting networks already deployed. Telecommunication networks have invested a great deal in third generation networks and there are several types of network to choose. This leads us here to suggest a new model for existing backbone providers of mobile networks: MANET, UMTS, WiMAX and satellite among others.

9.3. Demands of heterogeneous networks

The current infrastructure for telecommunication networks converges entirely on heterogeneous systems where users can be linked transparently to different networks during their communications. To ensure that there is an active connection in any service zone, simple switching mechanisms between different networks are needed. Moreover, a mobile terminal capable of functioning on several access technologies is therefore indispensable. Consequently, we can classify the demands of heterogeneous networks into two main groups:

– the first group involves operators to ensure interworking as well as to provide handover processes between the different access technologies;

– the second group involves the performance of user terminals to run and use network services.

9.3.1. *Demands on the network core*

Interworking is a term expressing the need to combine the advantages of different access technologies to obtain the best service. In a general sense, interworking means linking two or more distinct access networks to enable end users to access these networks and maintain service continuity. In heterogeneous networks, interworking between wireless broadband cellular networks can be carried out depending on two principal approaches: tight coupling and loose coupling [NGO 03].

In this section, we show an example of coupling in two UMTS and WiMAX networks.

9.3.1.1. *The "loose coupling" method*

This method calls on an essential gateway called a WiMAX gateway that interconnects the UMTS network to the WiMAX network. Figure 9.1 shows an example of interconnecting the UMTS network to the WiMAX network using the loose coupling method. In this interconnection, the gateway is connected to the Internet network via the AAA server of the UMTS network, without having any direct link to the latter.

Thus, the traffic from the WiMAX does not transmit via the UMTS core network, but passes directly to the Internet. We can say that the WiMAX network is a form of additional access network that offers its users direct access to the Internet. The loose coupling method of interconnection consists of interconnecting independent networks in the sense that they also provide independent services.

To manage authentication, invoicing and mobility in each network, several mechanisms are suggested. These mechanisms assume that WiMAX gateways support protocols ensuring mobility as well as the AAA server services relating to UMTS. This technique enables a UMTS network to collect the information necessary to manage a detailed invoice, by showing the usage and pricing for both networks.

Use of compatible AAA services enables WiMAX gateways to obtain, dynamically, service policies from local AAA servers. Thus, interconnection using loose coupling shows a number of assets. First, this approach ensures a degree of independence in the deployment and engineering of traffic. It also offers the UMTS network operator the advantage of deploying its network independently of owner–providers of WiMAX access networks and to benefit from the advantages of the latter with a minimum of investment.

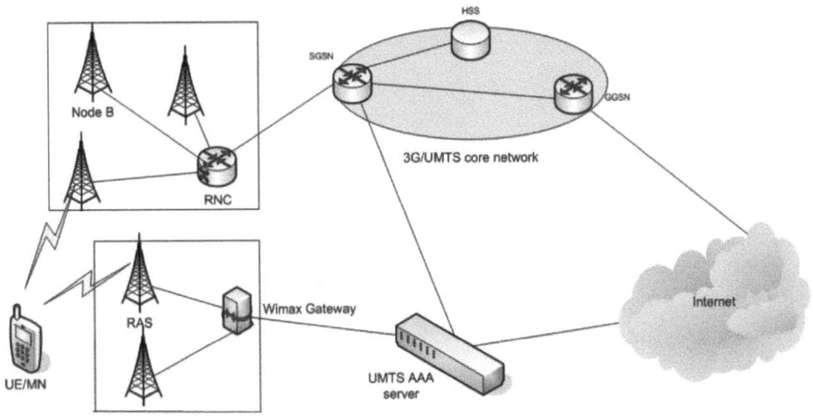

Figure 9.1. *Interconnection using the loose coupling method*

9.3.1.2. *The tight coupling method*

This method consists of interconnecting the WiMAX and UMTS networks so that the WiMAX network plays the role of another access network for the UMTS network. Thus, all the data traffic and signaling is channeled via the UMTS core network, with a modification of protocols to ensure the pre-requisite of interoperability.

In this interconnection approach, the WiMAX network is connected to the UMTS core network in the same way as any radio sub-system, running the features available in 3G RAN. This interconnection is carried out again by the WiMAX gateway, which makes all the downstream mechanisms transparent to the UMTS core network. Whatever the level of tight coupling, the data should imperatively transit via the UMTS core network before reaching the Internet network.

Interconnection using the tight coupling method can be done at three levels: GGSN, SGSN and RNC. At the GGSN level, we use equipment called VGSN (virtual GPRS support node). This equipment provides interfaces that serve to connect the WiMAX network at the WiMAX gateway on the one hand and the GGSN on the other hand, as illustrated in Figure 9.2. These act to exchange information on subscription and mobility, as well as the routing between the different integrated networks, which permits efficient management of each network's subscribers.

The VGSN can either be implemented as an independent stand-alone component, or integrated into the GGSN, or even integrated into the WiMAX gateway. We note that this configuration makes it possible to decrease congestion on the UMTS network, given that traffic bypasses a large part of this network.

However, the signaling traffic crosses the UMTS core network and requires very direct collaboration between the UMTS network and the WiMAX network. Thus, this technique remains efficient in terms of bandwidth per user as well as the duration of connections. We can, however, point out two major drawbacks to this technique: the increase in the delivery time if the position of the VGSN is not carefully chosen, and the risk of suffering the eventual consequences of breakdowns in the security and integrity of data [POC 99].

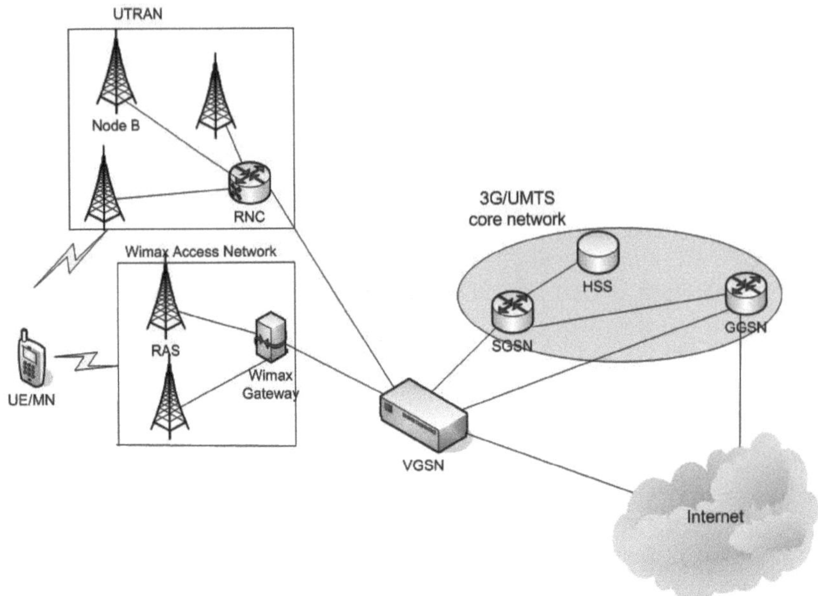

Figure 9.2. *Interconnection according to the tight coupling method at the GGSN*

In an interconnection using the tight coupling method at the SGSN, WiMAX is an alternative for subscribers who have access to a different access network as well as UTRAN. In other words, the UMTS core network does not differentiate between WiMAX radio technology and UTRAN. Figure 9.3 gives an example of interconnection using the tight coupling method at the SGSN.

The essential component in this coupling is the WiMAX gateway, which is connected to the SGSN and enables interconnection with the UMTS core network, while still concealing the peculiarities of WiMAX.

Interconnection using the tight coupling method can also be achieved at the RNC, as illustrated in Figure 9.4. In this approach, we note the existence of a component called IWU (interworking unit) making it possible to channel signaling exchanges and protocol conversions between the RNC and the different WiMAX base stations. This method offers a number of advantages. Among these, we mention efficiency in managing the mobility and continuity of services. Thus, users can maintain their active sessions while changing network. However, services remain subject to the demands on the Quality-of-Services (QoSs) on the WiMAX and UMTS network. We can still list the possibility of reusing the existing UMTS network, which would make it possible to make savings.

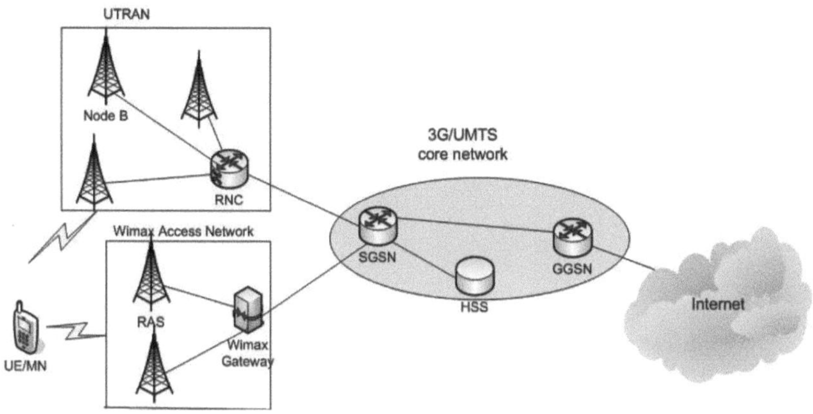

Figure 9.3. *Interconnection using the tight coupling method at the SGSN*

However, interconnection using the tight coupling method also has drawbacks. As an example, we cite the fact that the operator must be a global operator who must manage UMTS as well as WiMAX. If not, a user should have a subscription to both UMTS and WiMAX to ensure continuity of service [POC 99].

In conclusion, the distinction between tight coupling and loose coupling is based on the level of cooperation between the networks involved.

Tight coupling requires a major modification in the architecture of the access network and rewriting of the network protocol. The main advantage of loose coupling is that it does not require any architectural modification.

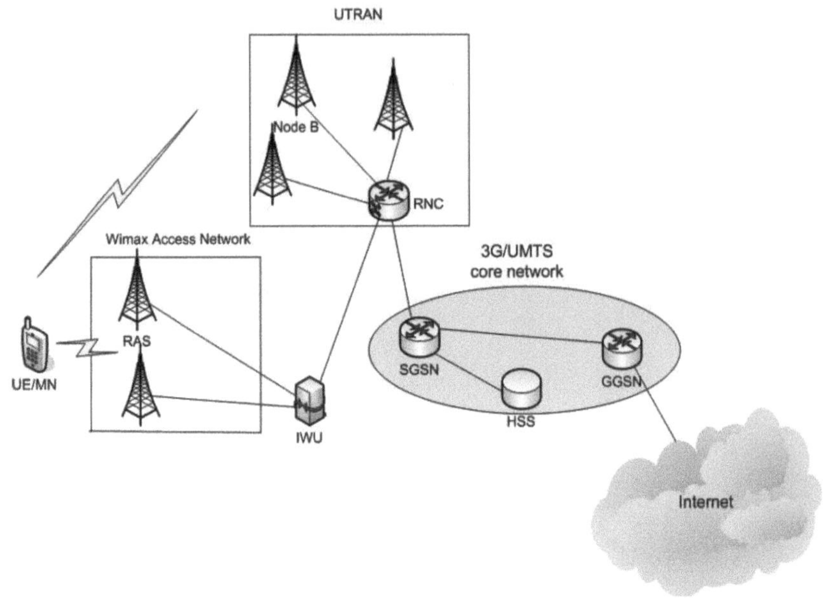

Figure 9.4. *Interconnection using the tight coupling method at the RNC*

However, agreements between operators who have different access networks are needed. Consequently, a user cannot maintain a connection between two access networks belonging to two non-collaborating operators, even if the user subscribes to both networks.

9.4. Other interconnection methods

9.4.1. *The interconnection suggested by the 3GPP project*

With the aim of ensuring interworking between wireless and mobile networks, the 3GPP project has defined in release 6 [100] the techniques for interworking between 3GPP cell systems with WLAN wireless networks. Then, in release 8, the 3GPP project also introduced an access network discovery and selection function (ANDSF) within the evolved packet core (EPC) network of the SAE (system architecture evolution) architecture to ensure interworking between 3GPP systems (such as 3G, HSPUS or LTE) and non-3GPP systems (such as WiFi and WiMAX). ANDSF provides user terminals with information on the available networks, constraints, rules and policies for inter-system mobility for each operator.

In conformity with 3GPP specifications, these data can be classified into three types depending on the operators' configurations:

1) Inter-system mobility policy (ISMP) to inform the terminal of the rules necessary for selecting and directing traffic on the single preferred access network.

2) Inter-system routing policy (ISRP) to specify rules at the terminal to select and direct traffic simultaneously on multiple access networks.

3) Access network discovery data to provide a list of networks available in the vicinity of the user with the aim of accelerating connection to these networks [RON 03]. In this way, the terminals are capable of quickly detecting the available access networks and taking decisions on the most appropriate choice of network for continuing the service.

9.4.2. *Interconnection suggested by the IEEE*

To contribute to the interconnection of heterogeneous networks, the IEEE has created a working group, IEEE 802.21 [WAT 89], with the aim of developing a norm to enable a mobile terminal to move transparently between different IEEE 802 access technologies. The norm is called media-independent handoff (MIH) services and was then extended to enable different cellular networks to integrate. It introduced a framework to support transparent transfers between different types of network [102].

Initially, the IEEE 802.21 working group focused on the following goals:

1) Creating terminals that are able to interact with different networks.

2) Respecting the principle of constantly active communication "always-on, anytime, anywhere, any networks" [ZHO 02].

3) Optimizing transitions across different networks so that they do not cut off or deteriorate: seamless mobility.

9.4.3. *The IEEE 802.21 standard*

The IEEE 802.21 standard consists of developing an architecture that enables transparent continuity of service when the mobile terminal moves between two heterogeneous networks at the data link level.

A set of functions enabling the handover to be optimized is defined in the mobility management (mobility management entity [MME]) protocol stack for elements of the network and a new entity called media-independent handover function (MIHF) is created.

Then, it consists of defining a service access point (MIH_SAP) and the associated primitives that provide MIH user with access to MIHF services.

Defining new service access points (SAPs) at the data link level and associated primitives for each access technology (IEEE 802.3, IEEE 802.11x, IEEE 802.16) is also taken into account by the IEEE 802.21 standard.

Moreover, the IEEE 802.21 standard specifies procedures that facilitate handover decision-makingand provide the state of the data link level in the form of information for MIH users. Thus, it provides the mobile station with the means to detect the networks available in its environment and enables it to initiate vertical handover between the networks detected. This makes it possible to carry out handovers with low latency between different access network technologies [WAT 89].

However, the norm simply makes it possible to provide information that aid handover initiation, network selection and activation of the interface. The norm does not include a definition of the handover policy and execution.

The approach adopted by the IEEE 802.21 group does not involve defining a complete solution but rather elements that can be integrated into an architecture for global interworking and global mobility. These functions support services facilitating the phases of access selection and preparation, and initiation for handover [ZHO 02].

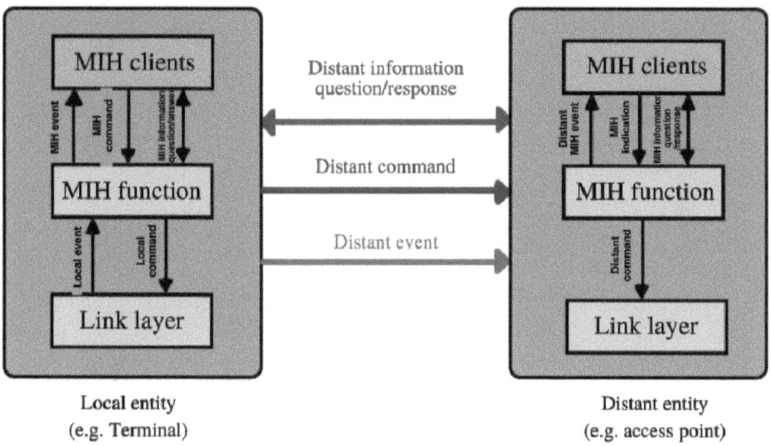

Figure 9.5. *MIH model of IEEE 802.21*

With the aim of ensuring correct performance of the MIH model, based on the following principles three services are supported:

– the MIH interfaces function with the two interlocutors' link layers (the access technology and the corresponding MIH). These client functions can, for example, be mobility protocols at the network level or higher such as MSCTP or SIP;

– the MIH functions can be distributed in the terminals and the network, and co-operate with each other to provide the required services;

– MIH services are used either locally in the node or at a distance between two MIH entities. The distant service makes it possible, for example, to propagate an event that has appeared in the terminal toward a network server or even enable network entities to control handovers;

– a mechanism for subscription to services with MIH functions is used to determine which events should be propagated either to an MIH client or to a distant MIH function.

The three supported services are as follows:

1) Event service (ES) to pick up the appearance of events that appear on the access interfaces.

2) Information service (IS) for discovering and exchanging information useful for the functions of network selection and handovers.

3) Command service (CS) for transmitting control commands.

To give a concrete example of these services, let us take the case of moving a subscriber from one network to another. The ES can pick up the appearance of a new radio access point, the IS can discover the name of the network and operator, and the CS can initiate handover to this new access point.

As we have already mentioned, MIH entities can be distributed in the network, for example, in the terminals, base stations and access points and mobility managers. This offers us more flexibility in the choice of installation of MIH bodies following certain criteria relating to mobility management and the type of network visited. On the other hand, MIH services can be used just as well in an approach where mobility is controlled locally at the level of the terminal or at distance in the network. The IEEE 802.21 standard only provides the global framework, whereas the algorithms that will really be implemented are left to designers.

Finally, in an approach comparable to MIH services, IEEE has also introduced norm 802.11u with a view to sustaining interworking of the WLAN networks with external networks [HUN 98]. IEE802.11u makes it possible to activate and automate network discovery and selection. Access points compatible with 802.11u give the mobile terminals information on the available networks likely to aid choice of the most suitable network and make adjustments necessary for the connection.

9.5. Transparency in communications

To ensure transparency in communications for mobile users passing from one network to another, two demands are indispensable for their mobile terminals.

The first concerns the terminal's ability to perform with all the available access networks. Meeting this objective requires integrating several network interfaces at each terminal or heterogeneous interfaces capable of adapting to several access technologies.

The second lies in the terminal's ability to collect the necessary information, process it and then take the decision to switch to the desired network.

9.5.1. *The process of decision-making for triggering an inter-network handover*

Management of inter-network or vertical handovers is a complex process based on decision policies for evaluating where and when the vertical handover will be triggered. This process unfolds in the following three stages:

– collection of information coming from the different networks present;

– the decision to trigger a vertical handover;

– carrying out the process of moving from one network to another.

In traditional approaches, the information collected to decide on the connection is linked to the received signal strength (RSS) [EKS 09]. On the contrary, in the case of a vertical handover, this selection strategy is not enough. As for the advent of new applications, users of new approaches to handover base their decisions on other information in addition to signal power.

Such a decision can be taken either at the level of the network, or at the level of the terminal. In this context, we distinguish four types of connection, discussed in the following sections.

9.5.2. *Mobile-controlled handover (MCHO)*

With this type of handover, it is only the mobile terminal that is responsible for the handover decision. It takes the necessary measures and combines the information required for the decision to be made. Then, based on the selection strategy considered, the mobile terminal takes the final decision. With the MCHO, the network takes no part in the decision. This type of handover has the advantage of reducing network complexity.

9.5.3. *Network-controlled handover (NCHO)*

Unlike a handover controlled by the mobile terminal, handover management of this type is carried out entirely by the network. Measurement of the parameters as well as the handover decision occurs in the network section. The mobile terminal takes no part in the handover decision.

9.5.4. *Mobile-assisted handover (MAHO)*

In this type of handover, the final decision is made by combining the efforts of the two parties involved: the mobile terminal and the network. In fact, it is the network that is concerned with taking the handover decision, but based on information and measurements received at the mobile terminal. This is the case with the GSM system.

9.5.5. *Network-assisted handover (NAHO)*

In the case of handover assisted by the network, the final handover decision is also taken because of the joint participation of the terminal and the network. But in this case, it is the network that combines the information needed for the handover decision and sends it to the mobile terminal to be used during the handover decision. Therefore, and unlike handover assisted by the mobile, it is the mobile terminal that has responsibility for taking the final decision using information sent from the part of the network concerned.

Execution of the process is dictated by a set of criteria that can be regrouped in the following way [BUD 02]:

– criteria for the network: coverage, latency, BER, etc;

– criteria for the terminal: speed, battery, information on location, etc;

– criteria for the user: user profile and preferences;

– criteria for services: QoS and potential services, etc.

9.6. Classes of service

Moreover, in a heterogeneous environment, the services can differ from one network to another. Currently, four classes of service can be distinguished: "conversational", "interactive", "streaming" and "background" [BUD 02].

9.6.1. *Conversational services*

Services in this class make it possible to transmit sound and images in real time, between two individuals or a group of people. This quality is therefore fairly sensitive to delays in transmission, as these delays are easily perceptible to users. Among these services, we can cite telephony and videophony [BUD 02]. The latter is less tolerant of errors than telephony and generally requires greater flow.

9.6.2. *Interactive classes of service*

The second class includes services requiring some interaction with the recipient in the form of requests/responses. An interactive application is characterized by the fact that the sender of the request awaits a response from the recipient for some time. Consequently, the

QoS is measured by the return time, that is, the time that passes between the moment when the message is sent and the moment when the response arrives. The recipient in interactive services can be a machine (e.g. a server) or a person.

9.6.3. *Services in the streaming class*

As for services in the third class, they use the technique of "streaming", which consists of sending a continuous flow of information and processing it instantaneously at the mobile terminal. The information used in these services is audio or video. Streaming services are used increasingly, as the majority of Internet users do not have fast enough access for downloading voluminous multimedia files in a reasonable time. With the help of the streaming technique, the client navigator can begin to display data before the entire file is downloaded. This class of services is one directional: transmission occurs only from the base station to the mobile user.

9.6.4. *Background class services*

Moreover, background services do not present, or scarcely present, any time constraint during transfer of the information that they generate. They can work in the "background".

The transfer time for these services can be measured in seconds, tens of seconds or minutes, without disrupting running of the user's other applications. Among these services are email transmission, SMS (short message service) and the download of database content.

9.7. Interconnection covering satellite, WiMAX, UMTS and MANET networks

In the previous chapters, we carried out a study of four different mobile networks: MANET for the category of mobile networks without fixed infrastructure; WiMAX and UMTS for the category of mobile networks with centralized network administration; and satellite networks combining DVB-S and RCS systems. We therefore started with MANET by suggesting a new extension to the SCTP protocol, which increases performances in less dense MANET. Then, we focused on suggesting models of interconnection between WiMAX mobile and UMTS, and WiMAX mobile and satellite.

Our aim in this section is to suggest a model for interconnection covering the four mobile and satellite networks described previously. This initiative belongs to the context of achieving seamless mobility. This concept enables a mobile subscription to move with complete transparency from one network to another without a break in connection and while respecting QoS.

To do this, we suggest three models for interconnection and compare the levels of mean output offered by each, with reference to simulations of a mobile station moving with a mean speed of 60 km/h.

9.7.1. *Interconnection models*

In this section, we describe the model for global network interconnection: DVB-S/RCS satellite, WiMAX mobile, UMTS and MANET are used for our comparative study. To do this, we rely on results obtained from previous simulations.

We begin our interconnection with the satellite network that uses the DVB-S/RCS norm and covers a cell with a radius of 50 km. The satellite responsible for diffusing the traffic to the MS in its coverage zone communicates directly with an NCC gateway on the ground; in turn it is connected to a router that ensures Internet access. This satellite network is interconnected to a WiMAX mobile network that covers a cell of 3 km. The intersection of these two cells forms a handover (HO) zone. For the WiMAX network, the BS is connected to an ASN-GW and this is linked to a CSN that ensures direct access to the Internet network.

On the other hand, the WiMAX network will be interconnected with a UMTS network, both forming an HO zone. UMTS covers a zone of 3 km. Node B is connected to the RNC that is linked to the SGSN, which is in turn connected to a GGSN; this ensures Internet access via the PDN network. The exchange of user data needed between the UMTS core network and the WiMAX's NSP is ensured by the WPDG gateway.

As for interconnection with the MANET network, this is formed of 10 mobile nodes. The first "source" node is found at the end of the UMTS coverage zone; it will be used to route from the MS to the "destination" node, which is directly connected to the Internet. Aside from the source and destination nodes, which are fixed, all the other nodes are mobile and continually change location.

9.7.2. *Simulation scenarios*

In this study, we focus on continuity in communication from the MS with its correspondent, during its passage from one network to another with a certain QoS. To do this, the MS should make several passages from the satellite network to the WiMAX, then the UMTS and finish by connecting to the CN via a MANET and vice versa throughout the duration of the simulation. Because the network resources are limited, to get closer to reality we install several other mobile stations in the satellite, UMTS and WiMAX coverage zones. These mobile stations begin to communicate with their respective network after 20 minutes from the start of the simulation. We also program a traffic-generating application between the MS and the CN to transit a maximum flow of 14,000 bit/sec. Using the results obtained from our previous simulations, the MS and CN will use CSTP for the satellite and WiMAX networks, MSCTP for the WiMAX and UMTS networks and SCTP–MANET for the MANET network.

With the aim of checking the qualitative contribution of our work, we carry out a comparative study between three different scenarios.

– *The first scenario*: A basic interconnection model – the signaling traffic and the vertical handover travel via the Internet (see Figure 9.6).

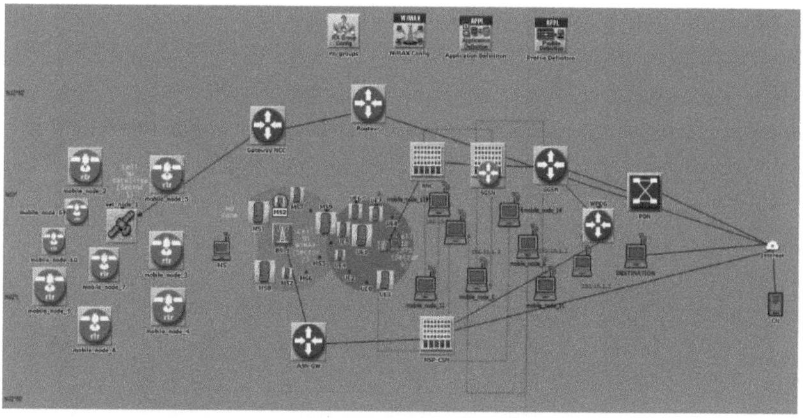

Figure 9.6. *Scenario 1: basic interconnection model*

– *Second scenario*: Interconnection model with SWU gateway – this will be connected to the NCC gateway of the satellite network on one side and to WPDG on the WiMAX and UMTS side to ensure transition of the signaling traffic (see Figure 9.7).

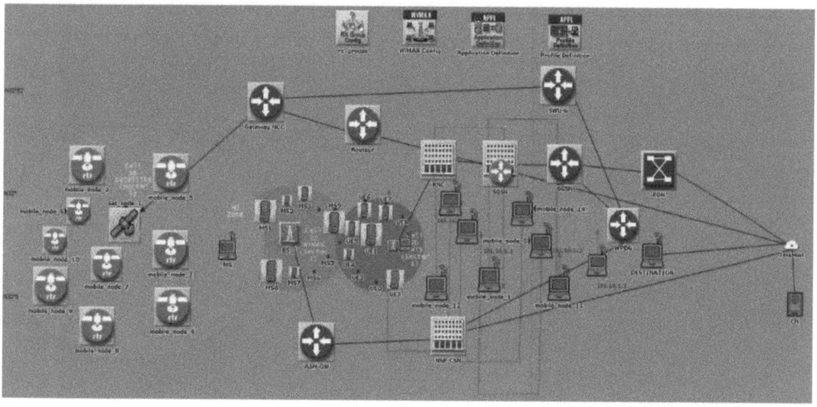

Figure 9.7. *Scenario 1: interconnection model with SWU gateway*

– *Third scenario*: Interconnection model with SWU gateway and MIH server – in this scenario, we have connected three MIH servers to the WiMAX satellite and UMTS networks and we have linked them together to an MIH server. Thus, the MS and the CN should install the MIH function. The SWU gateway is responsible for signaling and implementing vertical handover; the MIH server ensures detection of the type of network visited by the MS (see Figure 9.8).

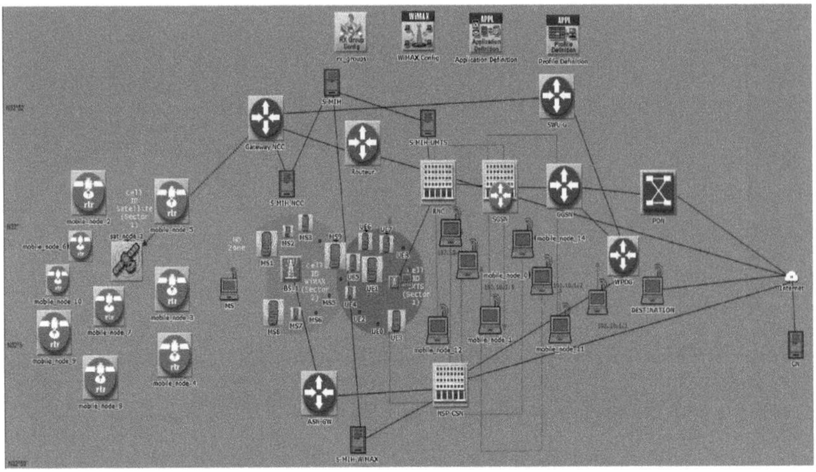

Figure 9.8. *Scenario 3: interconnection model with SWU gateway and MIH server*

9.7.3. *Simulation parameters*

For the networks' simulation parameters: satellites, WiMAX mobile, UMTS and MANET, we use those mentioned in the previous chapter.

9.7.4. *Performance criteria*

To compare the performances offered by the TCP, SCTP and MSCTP protocols in the case of vertical handover between satellite, WiMAX mobile, UMTS and MANET networks, we suggest the mean output received by the MS as performance criteria.

9.8. The results

In Figure 9.9, we observe the mean output received by the MS in bit/sec, depending on the networks' basic interconnection model: satellite, WiMAX mobile, UMTS and MANET compared to scenario 1. Thus, we can see that such an interconnection model offers a low output. The latter reaches a maximum peak of 11,000 bit/sec at the start of the simulation, to then drop below bar of 1,000 bit/sec at the end of the simulation.

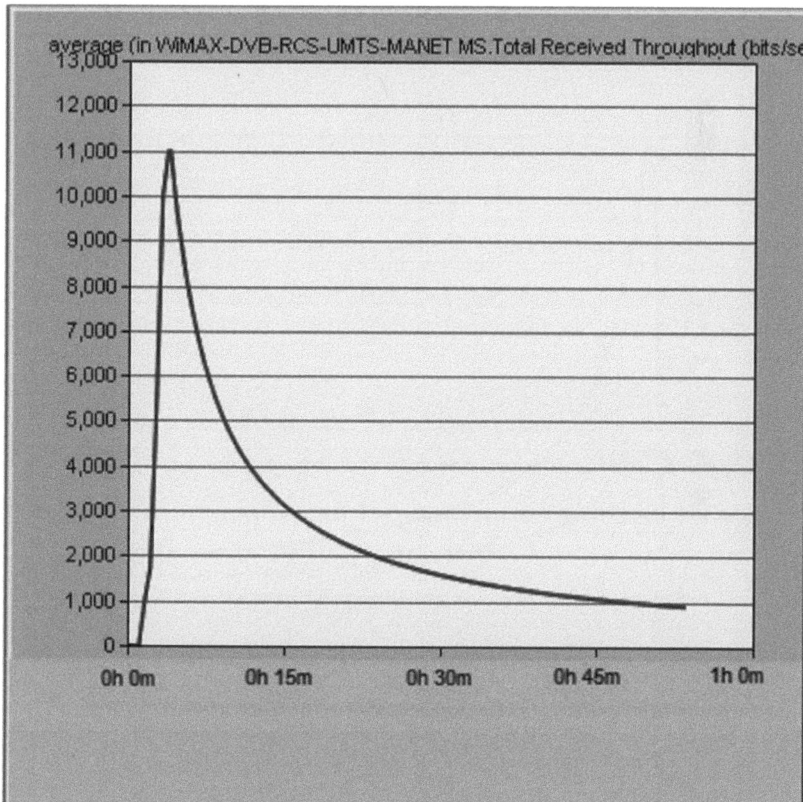

Figure 9.9. *Basic interconnection model*

Figure 9.10 shows a comparison between the mean output received by the MS compared to scenario 1 (basic interconnection model) and scenario 2 (interconnection model with SWU gateway). We note that with use of the SWU gateway the output does not improve as it reaches a maximum value of 3,892 bit/sec, but dissipates in a short duration of 8 minutes from the start of the simulation. Then the output decreases rapidly to drop below the bar of 500 bit/sec.

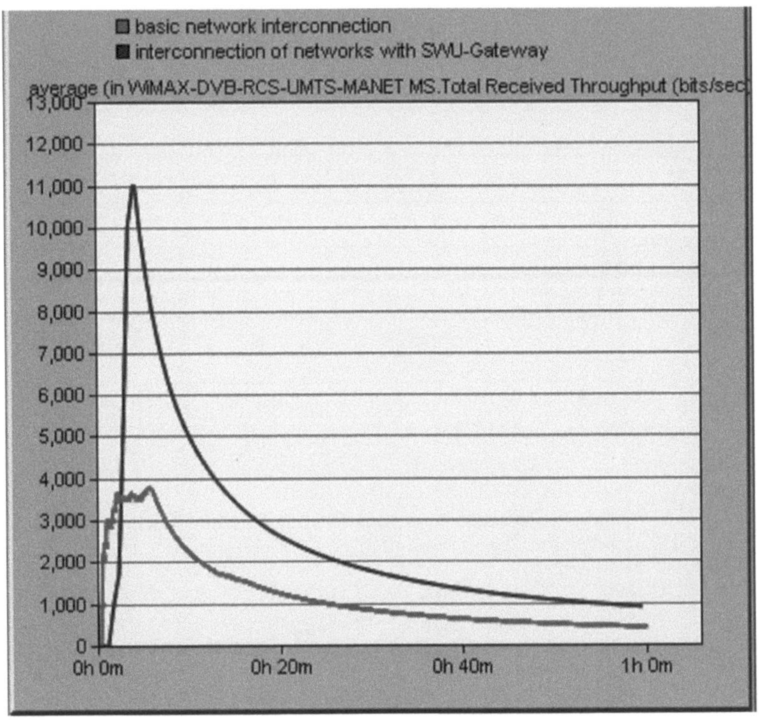

Figure 9.10. *Basic interconnection model and model with SWU gateway. For a color version of the figure, see www.iste.co.uk/benslama/networks.zip*

Figure 9.11 shows a comparison between the mean output received by the MS compared to scenario 1 (basic interconnection model) and scenario 2 (interconnection model with SWU gateway and interconnection with SWU gateway and MIH server). Even if the vertical handover mechanism between the different types of network is not totally transparent (performance oscillations), we note that by using the SWU gateway as well as an MIH server, the output improves significantly to reach 13,992 bit/sec at the start of the simulation, then it continues to oscillate between 9,000 and 6,000 bit/sec.

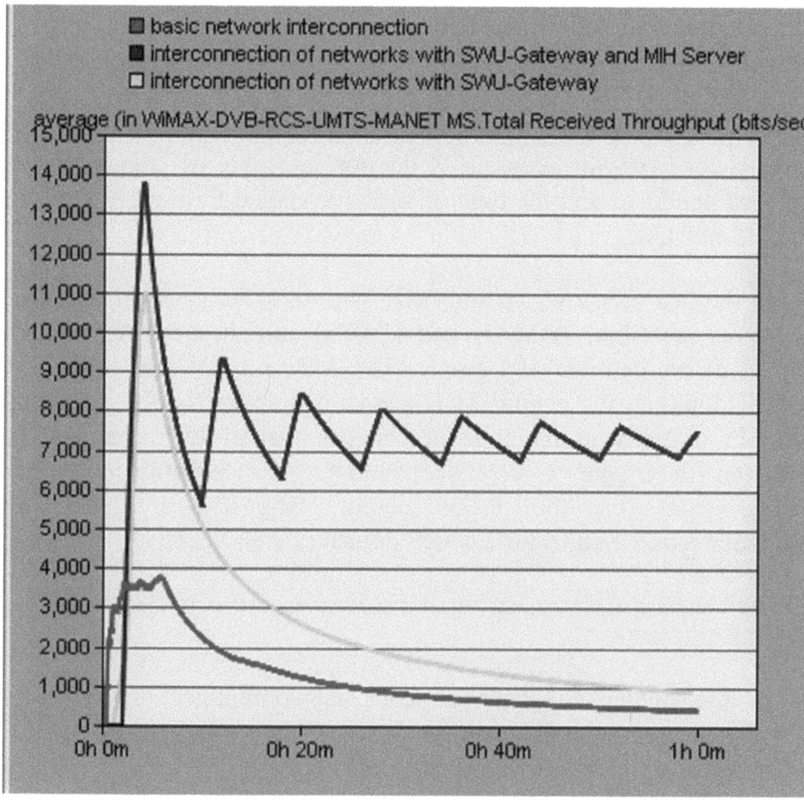

Figure 9.11. *Basic interconnection model and model with SWU gateway and MIH server. For a color version of the figure, see www.iste.co.uk/benslama/networks.zip*

9.9. Discussion of results

Creation of an interconnection model covers four networks: satellite, WiMAX mobile, UMTS and MANET, and in the context of global mobility, it is not based on the transport protocol alone, as the centralized mobility management architecture contributes to effective accomplishment of vertical handover.

We have demonstrated this through this comparative study between three interconnection models. In the first scenario, we recorded a very low output, which is explained by the transmission of signaling data and the triggering of vertical handover via the Internet (which causes substantial delays). To resolve this delay problem, we have suggested SWU gateway in scenario 2. The latter is responsible for the much quicker transfer of signaling traffic. Nevertheless, the results observed for scenario 2 do not improve, as trigger of the vertical handover and the type of network visited form a problem for the MS and CN.

In the last scenario, we add MIH servers at the level of the three networks (satellite, WiMAX and UMTS) and these servers are all linked to the central MIH server. The MS and CN also install the MIHP function; the central MIH server is responsible for informing the MS and CN of the type of network visited and promotes the decision to trigger a vertical handover. Consequently, this has repercussions on the mean output suggested by such an interconnection architecture, which improves significantly.

9.10. Conclusion

In the domain of heterogeneous telecommunications, a large number of users with different types of terminal, and different choices of preference and metric values are expected.

Therefore, control of handover by the network alone or with the help of information received from the mobile terminals becomes very difficult and increases complexity at the level of the network. Moreover, information on the precise parameters of the networks available are indispensable. Consequently, the vertical handover process in the heterogeneous domain should be controlled by mobile terminal and supported by the participating networks. In the context of achieving global mobility by integrating the concept of multihoming, we suggest concentrating on two essential points:

1) The interconnection model that is responsible for defining the type of network visited; it thus promotes the decision to trigger a vertical handover.

2) The transport protocol that ensures an optimal output depending on the constraints of speed and the number of users.

10

Compressed Sensing in Radar Interferometry

Compressed sensing is a new framework for sampling and reconstructing sparse signals using small enough measures. This new strategy has been applied to synthetic aperture radar (SAR) and it links two principles: sparsity, which is linked to the signal we are interested in, and incoherence, which refers to methods of sensing. An important question in designing SAR systems based on compressed sensing concerns the necessary, sufficient sampling rate for an exact reconstruction of sparse signals. The measurement limits or sampling rates in compressed sensing have been taken from information theory. This additional theoretical information should be validated in the context of SAR imaging. Zhang offers an analysis of the sub-Nyquist sample rate in the case of SAR [ZHA 14]. A second point deserves our attention, which is the choice of norms, which is an essential element in processing compressed sensing. We discuss these points in this chapter; we present the problem of subsampling in SAR inferogram processing, then that of choosing norms in processing maps.

10.1. Formatting inferograms

Now that the principles of interferometry have been introduced, we discuss certain practical aspects of creating inferograms, especially the way they are produced by CNES [MAS 92, MAS 93a]. First, we introduce registration techniques and the use of geometric information that results from them. In addition to the strictly defined inferogram, this stage makes it possible to calculate the complex correlation between the two initial SWR images, including the module, called coherence, and makes it possible to characterize the phase's validity. We specify the way in which these two data are estimated by taking account of the orbital fringes and by reducing noise by a consistent averaging called "complex multi-looking".

10.1.1. *Registration*

The processes carried out on SWR data to obtain the inferogram are complex and require substantial expertise in the domain of radar as well as in the domain of images. The techniques used differ depending on the laboratories and the nature of the initial data. They are rarely explained in detail because of the issues related to mastering this first stage in the chain of inferometric processing.

Registration conditions the quality of the data as well as their exact interpretation.

We therefore mainly detail the initial processes for SWR images. This process includes three stages:

1) The formation of two images that should take account of the constraints of inferometry.

2) The determination of geometric transformation, making it possible to calculate one homologous point in one image to that in another.

3) Re-sampling the registered image onto the master image without the need to introduce a phase-encoded artifact.

10.2. SWR synthesis

The first stage is to reconstruct the two complex images. In the first place, the SWR correlator must preserve the phase information, which is not always the case. Prati [PRA 89, PRA 90] and Madsen [MAD

93] suggest synthesis algorithms for interferometry. CNES has also developed an SWR correlator [MAS 93b] guaranteeing the performances required for interferometry.

Moreover, it is necessary for the resolution cells to be seen in both images at the same angle of view to guarantee that the phase is coherent. We have seen in the introduction to M-views processing that this angle depends on the central Doppler frequency used during azimuth processing. Knowing the parameters of the synthesis of the first image, it makes sense to adapt the central Doppler frequency used during synthesis of the second in such a way as to obtain the desired angle of view.

These corrections are indispensable in the case of non-parallel orbits where it is necessary to artificially straighten the trajectories. They have been applied successfully to SIR-B data by teams from JPL (Jet Propulsion Laboratory NASA USA) and CNES *Centre National (d'études Spatiales France* [GAB 88] and CNES [MAS 93c].

In the case of almost parallel trajectories such as those of ERS-1, images should theoretically be synthesized with the same central Doppler frequency. To gain time and data volume, Massonnet suggests shaping the inferogram directly by synthesizing the two images simultaneously.

10.3. Geometric deformation

The following stage consists of searching for the geometric transformation making it possible to move from the slave image to the master image. The most classical methods are based on correlation techniques. If we write the two images' radiometry as ρ_A and ρ_B, then for a pixel window $(i,j) \in F$, we seek the parameters of the relative movement (k,l), which maximizes the intercorrelation function:

$$C(k,l)=\frac{\sum_F\left(\rho_A(i,j)-\mu_A\right)\left(\rho_B(i+k,j+l)-\mu_B\right)}{\sqrt{\sum_F\left(\rho_A(i,j)-\mu_A\right)^2\cdot\sum_F\left(\rho_B(i+k,j+l)-\mu_B\right)^2}} \qquad [10.1]$$

where μ_A and μ_B are the averages of the radiometries of A and B, respectively, over F. This method has been used by Li [LI 90]. The discreet values of $C(k,l)$ are then interpolated in the vicinity of the maximum to obtain a registration with less precision than the pixel. This correlation between intensities is also used to obtain an initial registration between the two images. However, the irregularity of the radiometry of SWR images limits the quality that can be expected from this method. It is necessary to refine the registration with the help of other methods specific to inferometry.

In order to carry out with precision the important registrations necessitated by SIR-B data, Gabriel has developed a method based on the presence of strong local frequencies in the inferogram [GAB 88]. When the registration is good, the spectral power of a reduced window should concentrate around the local frequency of the fringes.

The method consists of moving an 8 × 8 vignette of the slave image into the master image and calculating the spectrum $F(f_x,f_y)$ (two-dimensional [2D] Fourier transform) of the 8 × 8 inferogram obtained in different positions. As a measure of the quality of the registration, they use the spectrum's signal-to-noise (SNR) ratio defined by:

$$SNR = \frac{\max\limits_{(f_x,f_y)} \left| F(f_x,f_y) \right|}{\sum\limits_{(f_x,f_y)} \left| F(f_x,f_y) \right|} \qquad [10.2]$$

As before, interpolation of the values obtained in the vicinity of the maximum permits a subpixel registration of the 8 × 8 window. This method thus makes it possible to place a certain number of anchor points between which the deformation is calculated by linear regression. The drawback of this technique is the very high calculation time.

Prati initially used a very similar method, five anchor points with the help of the same SNR, after having oversampled the two images by a factor of 8 [PRA 89]. This technique should take the orbital fringes into account as soon as they are known, to reduce loss due to the phase's natural evolution.

Another technique from inferometry is suggested by Lin [LIN 91] to determine the three registration parameters: the offset in azimuth k, the offset in distance l and the scale factor in distance h. After oversampling the slave image by a factor of 10, Lin calculates for different values of the parameters a global cost given by the mean fluctuation function defined on the inferogram by:

$$C(k,k,h)=\sum_i\sum_j\left(\left|\phi(i+1,j)-\phi(i,j)\right|+\left|\phi(i,j+1)-\phi(i,j)\right|\right) \quad [10.3]$$

This cost relies on the hypothesis of a slow variation in the inferometric phase when it is only a function of the difference in path. Lin indicates that when the registration is good, the fringe edges are very well defined and the $0 \leftrightarrow 2\pi$ transitions have a minimum contribution. Inversely, when the registration is bad, the transitions are "thick" and influence a large number of pixels.

10.4. Re-sampling

We have just seen that several of these methods require oversampling of one or both initial images. In addition, once the geometric deformation is found, we need to be able to estimate the complex value of the slave image at point $P_e(x,y)$ corresponding to pixel $P_m(x,y)$ of the master image. Coordinates x and y no longer belong to the discreet grid of the slave image and this value should be calculated from the neighbor pixels of $P_e(x,y)$. In both these cases, we stumble on the problem of re-sampling a complex image and especially on that of determining the phase. A well-known defect of this type of manipulation on intensity images is the modification of the data's statistical properties [QUE 89].

Classical linear interpolation techniques (bilinear in the case of 2D) or at higher degrees (quadratics, splines, etc.) are generally aimed at real signals. They are nevertheless applied by Gabriel [GAB 88] and Lin [LIN 91], working separately on real and imaginary parts.

A routine technique for oversampling a signal is to periodize the spectrum issuing from the fast Fourier transforms (FFTs 1D or 2D). It is used by Prati for oversampling the two images by a factor of 8 [PRA 89], as well as increasing the distance sampling by a factor of 2. The latter however calculates the re-sampled image following a subpixel convolutional recoding in the space domain with the help of the cardinal sines in both directions.

This leaves re-sampling of the slave image using the nearest neighbor technique. It offers the advantage of not introducing new values inconsistent with the unpredictable speckle phenomenon that underlines the values of SWR images.

10.5. Geometric interpretation of the registration

If both images have been synthesized correctly and obtained on parallel orbits, they are in geometry comparable to epipolar geometry in stereo optical images. The azimuth recoding should be obtainable by a global translation. This does not contribute any information on the relief.

Distance is however the first stage in estimating the relief via the difference in pathway. If we return to the path difference calculation in equation [10.1] by introducing n_A (respectively, n_B), the position of a target in image A (respectively, B), counted with the pixel size in distance as a unit, we have:

$$n_A = \frac{1}{r}\left(\sqrt{x^2 + (H-z)^2} - R_A^0\right) \qquad [10.4.a]$$

$$n_B = \frac{1}{r}\left(\sqrt{(x-b)^2 + (H-z)^2} - R_B^0\right) \qquad [10.4.b]$$

where $R^0{}_A$ (respectively, $R^0{}_B$) represents the proximal range, that is, the target radar distance from the start of the sampling at distance. By returning to parameter $d = \sqrt{x^2 + H^2}$, the difference in distance localization is then:

$$\delta n = n_A - n_B = \frac{1}{r}\left[\left(R_B^0 - R_A^0\right) + \frac{xb}{d} + \frac{zxbH}{d^3}\right] \quad [10.5]$$

We can interpret the three terms of this equation in the following way:

– The first term depends only on the proximal distances. It corresponds to the registration between the beginnings of the line. Carried out, for example, by correlation between the sizes, this registration can lack precision (in the order of 0.1 pixel). The error introduced on the difference in size is inconsequential as it is only introduced once and the measurement for the differences in distance is then counted in relation to this origin. Then we consider this termas null, which amounts to considering the two-line beginnings already registered.

– The second and third terms are functions of the target's position in distance on the ground x and at altitude z. They correspond, respectively, to the orbital and topographical fringes. The ratio between these two terms (equation [10.5]) highlights the preponderance of the orbital term.

Taking account of the registration's weak precision, of some hundreds of pixels, we can consider that the distortion in distance provided by the registration measures only the second term. In other words, for the small bases used in inferometry, registration (the classical stereo) does not make it possible to reach the relief, but provides an estimation of the distortion in distance due to that distance. This estimation will be usable to "roughly" eliminate the orbital fringes and facilitate use of the inferogram.

10.5.1. *Resulting images*

Once the two images are registered, in each pixel (i,j) we use both complex values, $S_A(i,j)$ and $S_B(i,j)$. The inferometric phase can be calculated as the argument of the product:

$$S_A(i,j).S_B^*(i,j) = \rho_A(i,j)\,\rho_B(i,j)\,e^{j(\phi_A(i,j)-\phi_B(i,j))}$$

But this estimation has a lot of noise. To diminish its variance, we are forced to sacrifice the spatial resolution by carrying out a consistent averaging of M products $S_A(i,j)\ .\ S^*_B(i,j)$. These M values are obtained either by a true synthesis of M, seen from each of the two images, or by a spatial averaging of N pixels of a window F taken on the monovision images A and B. In this case, M is the number of equivalent views [ZEB 94] defined by the number of initial resolution cells contained on the ground surface s_p linked to the new pixel P:

$$M = \frac{s_P}{\mathrm{Re}s_x\,\mathrm{Re}s_y} \qquad [10.6]$$

The phase obtained at pixel P is then given by the argument of the complex number $Z_0(P)$:

$$Z_0(P) = \frac{\sum_{(i,j)\in F} \rho_A(i,j)e^{j\phi_A(i,j)}\rho_B(i,j)e^{-j\phi_B(i,j)}}{\sqrt{\sum_{(i,j)\in F}\rho_A^2(i,j)}\cdot\sqrt{\sum_{(i,j)\in F}\rho_B^2(i,j)}} \qquad [10.7]$$

Module $|\,Z_0(P)\,|$ is a measurement of the complex correlation between the two images.

10.5.2. *Eliminating orbital fringes*

However, this spatial averaging assumes that the inferometric phase remains constant on window F. Indeed, the phase turns because of the topography and the slant distance. If the first factor is initially inaccessible, the main contribution of the second can be approximated, either with the help of orbitographic data, or intrinsically by estimating the main frequency in distance on the initial inferogram or using the geometric information resulting from the registration.

Consider equation [10.5]: we write as $\delta\tilde{n}(i,j)$ the correction in distance measured in pixels after registering the beginnings of the lines. If Q is the number of wave lengths λ by pixel at distance:

$$Q = \frac{2r}{\lambda} = \frac{T_e}{T_0} = \frac{f_0}{f_e} \qquad [10.8]$$

where f_e is the sampling frequency of the return and f_0 the radar carrier frequency (T_e and T_0 and the corresponding periods). For ERS-1 images, we have: $Q = 279{,}495 \pm 0.01$.

The inferometric phase breaks down into two terms:

$$\Phi(i,j) = \Phi_{orbital}(i,j) + \Phi_{topographic}(i,j) \qquad [10.9]$$

the first term can be estimated using registration by:

$$\emptyset_{orbital}(i,j) = 2\pi Q\tilde{n}(i,j) \qquad [10.10]$$

Massonnet suggests modeling the distortion $\delta\tilde{n}(i,j)$ using a second-degree 2D polynomial [MAS 92].

This estimation is then used simultaneously to subtract the orbital part from the phase difference and to reduce noise using "complex multi-looking":

$$Z(P) = \frac{\sum_{(i,j)\in F} S_A(i,j).S_B^*(i,j)e^{-j2\pi Q\delta\tilde{n}(i,j)}}{\sqrt{\sum_{(i,j)\in F} \rho_A^2(i,j)}.\sqrt{\sum_{(i,j)\in F} \rho_B^2(i,j)}} \qquad [10.11]$$

The inferograms' phase $\Phi(P)$ to which we will refer is then defined by the argument:

$$\phi(P) = \arg\left(\sum_{(i,j)\in F} S_A(i,j).S_B^*(i,j)e^{-j2\pi Q\delta\tilde{n}(i,j)}\right) \qquad [10.12]$$

Similarly, the correlation $\gamma(P)$ to which we will then refer is given by the module:

$$\gamma(P) = \left|\frac{\sum_{(i,j)\in F} S_A(i,j).S_B^*(i,j)e^{-j2\pi Q\delta\tilde{n}(i,j)}}{\sqrt{\sum_{(i,j)\in F} \rho_A^2(i,j)}.\sqrt{\sum_{(i,j)\in F} \rho_B^2(i,j)}}\right| \qquad [10.13]$$

This spatial averaging is more "proper" than the previous one (equation [10.7]), with the reservation that the topographical fringes remain broad enough in the inferogram resulting from weak resolutions in both directions. An order of size of the moving average window is 2 pixels in distance and 10 in azimuth, which gives a resulting pixel of approximately 40 m × 40 m on the ground.

10.6. Simulated inferograms

To compare filtering efficiency, we have chosen two widely used spatial filters: morphological filtering and median filtering. For the performance, we have chosen the Goldstein method, whose flooding occurs by simple integration. The criteria are as follows: image quality, the surface of the distorted region, the root mean square error (RMSE), the rates of residue reduction and the mean square deviation (MSD), the regions are masked by the branches and computation time.

RMSE is the error between the original image and unwrapped image, so a minimum RMSE algorithm is better:

$$\mathrm{RMSE} = \sqrt{\frac{\sum_{ij=1,1}^{MN}(\psi_{i,j}-\psi_{i,j}^{0})^2}{MN}} \qquad [10.14]$$

MSD is one of the reliable criteria for the test; it is used to compare the phase deviation between the unwrapped image and the original image to validate the efficiency of one method for the profile of another [SHI 08].

$$\mathrm{MSD} = \sqrt{\frac{1}{\mathrm{M.N}}\sum_{\mathrm{i}=1}^{\mathrm{M}}\sum_{\mathrm{j}=1}^{\mathrm{N}}\left(\psi - \frac{1}{\mathrm{M.N}}\sum_{\mathrm{i}=1}^{\mathrm{M}}\sum_{\mathrm{j}=1}^{\mathrm{N}}(\psi - \psi_{\mathrm{o}})\right)^2} \qquad [10.15]$$

If ψ_0 is the original phase and ψ is the unwrapped phase, then the algorithm with minimum MSD is better.

The simultaneous inferograms chosen are as follows:

1) 100 × 100 pixel images of peaks with a low level of closed fringes (Figures 10.1(a–c)).

2) 126 × 126 pixel image of waves with a high rate of open fringes (Figures 10.2(a–c)).

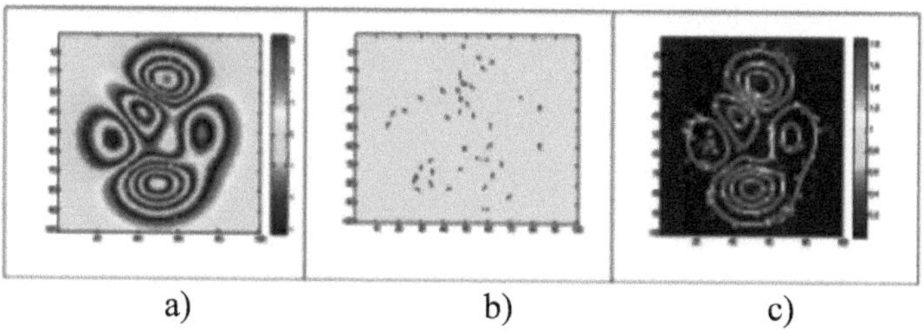

Figure 10.1. *Test with images of peaks. a) Noisy inferogram, b) map of residue before filtering, c) PDV map. For a color version of the figure, see www.iste.co.uk/benslama/networks.zip*

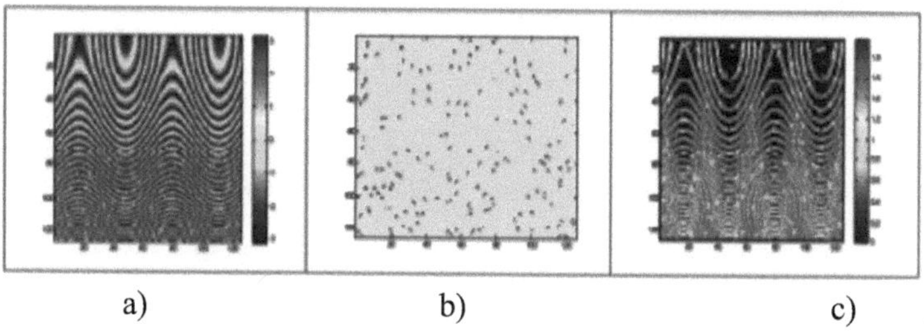

Figure 10.2. *Test with images of waves. a) Noisy inferogram, b) map of residue before filtering, c) PDV map. For a color version of the figure, see www.iste.co.uk/benslama/networks.zip*

Although the noise (the discontinuities) is generated by the same random processes and with the same density, the amount of residue in the image with peaks (Figure 10.2(b)) is smaller than in the image of waves (Figure 10.2(c)) because the image with peaks has fewer fringes than that with waves, so filters based on smoothing, averaging, dilation or erosion can give unsatisfactory results in terms of residue reduction if the number of fringes is substantial.

Figures 10.3(b) and (c) show that the PDV function is likely to clarify clearly those fringes whose values are between 1 and 1.8 (see the color bar in the figure). Consequently, the suggested filter, using this function as a guide to quality, can give good results to avoid smoothing fringes with a minimized time in relation to the directional masks, it can estimate the poor pixels among each of the adjacent 4, since the minimized PAD filtering center is only those pixels selected.

Qualitative results are shown in Figures 10.3, 10.4 and 10.5.

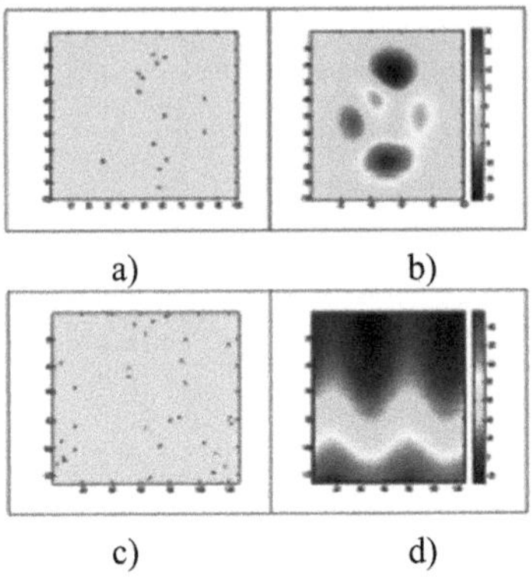

Figure 10.3. *Filtering by PDV-PAD. a) Map of residue after filtering (peaks). b) Unwrapped phase image (peaks). c) Map of residue after filtering (waves). d) Unwrapped phase image (waves). For a color version of the figure, see www.iste.co.uk/benslama/networks.zip*

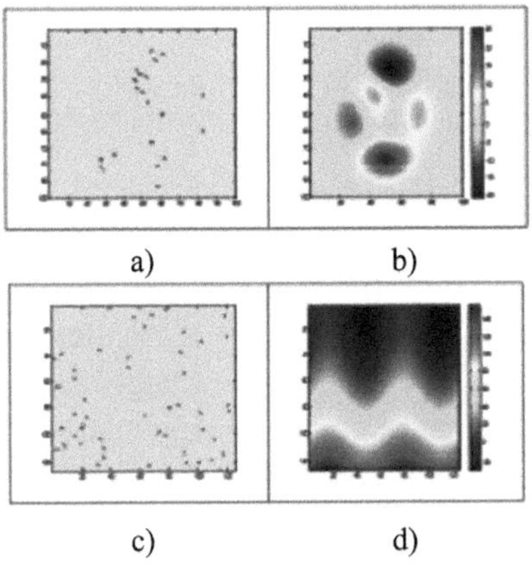

Figure 10.4. *Filtering using morphology. a) Map of residue after filtering (peaks). b) Unwrapped phase image (peaks). c) Unwrapped phase image (waves). d) Unwrapped phase image (waves). For a color version of the figure, see www.iste.co.uk/benslama/networks.zip*

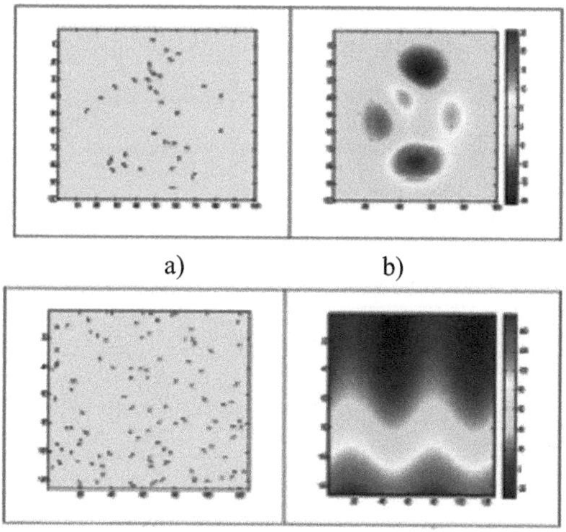

a) b)

c) d)

Figure 10.5. *Filtering using the modified median. a) Unwrapped phase image (peaks). b) Unwrapped phase image (peaks). c) Unwrapped phase image (waves). d) Unwrapped phase image (waves). For a color version of the figure, see www.iste.co.uk/benslama/networks.zip*

10.6.1. *Performances in terms of distorted regions*

Figure 10.6 shows performances of the suggested adjustment method and that of simple integration in terms of the surface of the distorted region (number of pixels corresponding to discontinuities). To this end, we have voluntarily omitted two inverse sine residues outside the branches (unmasked); these pixels are found in a diagonal position at the center of the inferogram and are gradually separated, therefore this curve is a function of the distance separating both unmasked residues. Figure 10.7 shows an example of a diagonal distance of 3 pixels.

According to this curve, the surface of the distorted region after operating is very small using the adjustment method compared to the simple integration method. Effectively, with regard to the mean of all the distances tested, it is only 23.1667 for the suggested method against 178.5 for the SI method; it is a mean difference of 155.33 pixels.

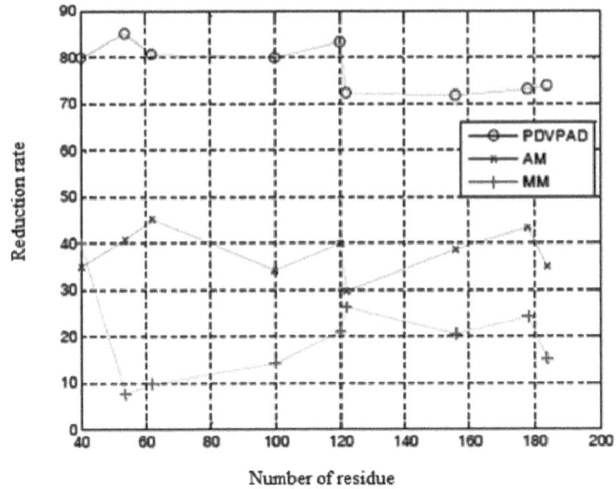

Figure 10.6. *Operating performances suggested in terms of the surface of the distorted regions. AJ, adjustment; SI, simple integration. For a color version of the figure, see www.iste.co.uk/benslama/networks.zip*

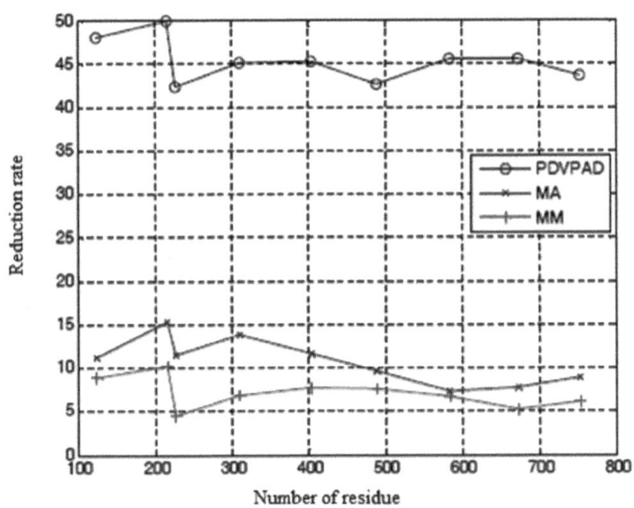

Figure 10.7. *An example of a diagonal distance separating two unmasked inverse sine residues. For a color version of the figure, see www.iste.co.uk/benslama/networks.zip*

10.6.2. *Performances in terms of RMSE*

Figure 10.8 evaluates performances in terms of RMSE (the square of the quadratic mean error) between the phase unwrapped by the AJ/SI method and the original image for the same distances as in Figure 10.7.

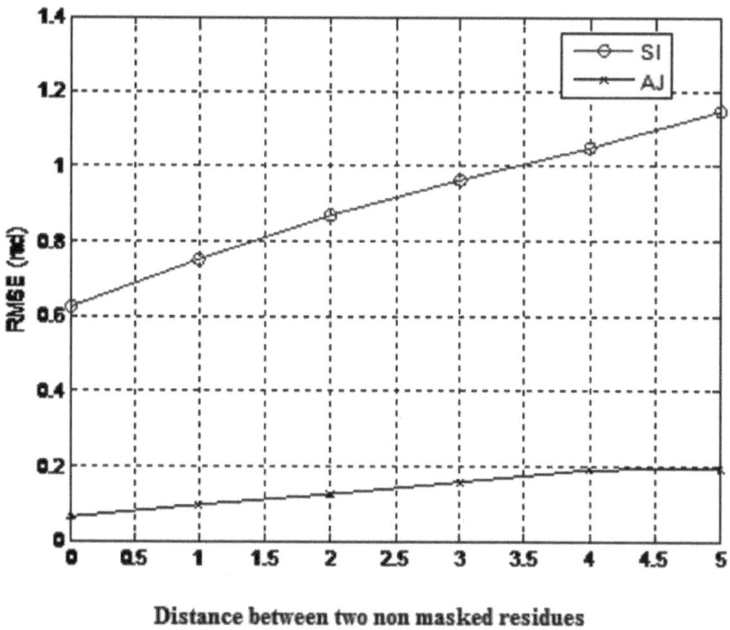

Figure 10.8. *Operating performances suggested in terms of RMSE*

Mean SI = 0.9020, mean AJ = 0.1374, which is a mean difference of 0.7646 rad.

Similarly, the suggested operating shows small values of RMSE for all the distances chosen. On average, SI = 0.9020, AJ = 0.1374, which is a difference of 0.7646 rad. This confirms the limitation of error propagation by applying the adjustment method in case of residues poorly masked by the procedure for connecting branches.

10.7. Methods for minimizing norms

In the previous section, we introduced a state-of-the-art on operating methods in the path following category – which are the most well-known and are used as benchmarks. Now, we turn to the second category: minimal-norms (MN). This involves other approaches that rely on a completely different concept based on function minimizations. Although path following (PF) methods have a local vision, that of MN has a global operating vision, treating the problem with optimization procedures, which involve the entire inferogram at once.

10.7.1. *L*^{p} *norms*

MN methods try to find the solution of the absolute (unwrapped) phase image ψ for which the L^p norm of the difference (horizontal and vertical) between the offset of the unwrapped phase and offset of the wrapped phase is minimized; as we have explained earlier, the wrapped and unwrapped phase gradients should be unchanged for a case without error.

This type of method seeks wrapped phase images whose local offset corresponds to the offset of the wrapped phase. This minimization aims to find the solution $\hat{f}$, which can in general be expressed by:

$$\hat{f} = argmin\, E(f) \qquad [10.16]$$

Argmin means the minimization argument, *E*(*f*) is the L^p norm used for minmization, which is:

$$E(f) = \mathrm{i} \sum_{i=1}^{M} \sum_{j=1}^{N-1} \left|\Delta^h \phi_{i,j} - \Delta^h \Psi_{i,j}\right|^p + \sum_{i=1}^{M-1} \sum_{j=1}^{N} \left|\Delta^v \phi_{i,j} - \Delta^v \Psi_{i,j}\right|^p, p \geq 0 \qquad [10.17]$$

with:

$$\Delta^h \phi_{i,j} = \phi_{i,j+1} - \phi_{i,j}, \Delta^v \phi_{i,j} = \phi_{i+1,j} - \phi_{i,j}$$

$$\Delta^h \Psi_{i,j} = w\left(\Psi_{i,j+1} - \Psi_{i,j}\right), \Delta^v \Psi_{i,j} = w(\Psi_{i+1,j} - \Psi_{i,j})$$

Therefore, solution $\hat{f}$ leads to a function f, which is the possible wrapped phase (ψ); in addition, the difference between the terms ϕ and ψ is the entire k, this is another possible way of applying the differences without wrapping phase ψ.

We can also weight this approach if we have a map of quality, Q by determining this map both horizontally and vertically, Q^h and Q^v [67].

So equation [10.17] can be written as:

$$E(f) = \sum_{i=1}^{M}\sum_{j=1}^{N-1} q_{i,j}^{h}\left|\Delta^{h}\phi_{i,j} - \Delta^{h}\Psi_{i,j}\right|^{p} + \sum_{i=1}^{M-1}\sum_{j=1}^{N} q_{i,j}^{v}\left|\Delta^{v}\phi_{i,j} - \Delta^{v}\Psi_{i,j}\right|^{p}, p \geq 0 \qquad [10.18]$$

The map of quality can be divided by a threshold and consequently it becomes binary, so in the noisy regions with residue, the parameter $q_{i,j}$ is equal to 0, and the good regions are weighted by 1.

These expressions [10.17] and [10.18] show the global aspect of this approach; the fact that all the phases are observed is used to calculate the solution. The different values of p lead to different variants of this approach. Generally, the values $p \leq 2$ are used and often cited in literature.

10.7.2. L^2 norm

With $p = 2$ we have the least squares problem. In this case, we note that the minimization, even for the square, is a discreet minimization that is much in demand for recent suggestions, yet the majority of the algorithms are approximate solutions; one of the drawbacks of norm L^2 is that it tries to smooth the discontinuities unless it is used with binary weighting.

Fried and Hudgin were the first to suggest an unwrapping (with pre-filtering) using least squares approximation [FRI 77, HUD 77]. Then several algorithms were published; some relax the calculation procedure from Z to R because the discreet nature of the problem is more complex. In the continuous domain, the problem is equivalent to solving Poisson's partial differential equation [GHI 98]. This can be solved using the FFT or DCT [SHI 10, GHI 94], then returning to the discreet domain [EGI 04, WAN 14, ZHA 14]. An exact solution for the least squares method has been developed in [DIA 02] using network programming techniques, and in [DIA 07] using the max flow – minus discontinuity.

10.7.3. *L^1 norm*

The L^1 norm is better than the L^2 norm in terms of preserving fringes as L^2 smoothes the discontinuities. This method has been suggested by Flynn [FLY 97] and Costantini [COS 98, COS 99] using the network programming technique, and by [WAN 14] using max flow in L^2 norm and PDV as a map of quality.

10.7.4. *Small value norms of p*

With $0 \leq p < 1$, the ability to preserve the discontinuities is better ameliorated, but with a complexity that should not be dismissed for this type of norm. In particular, L^0, which is very complex but on the other hand very desirable in practice [CHE 00].

10.8. Conclusion

In this chapter, an introduction to the phase sequence problem, relevant formulations, a state of the art on some major methods, analysis and comparison have been explored. This has made it possible to carry out a sufficient analysis to understand the mechanisms behind each approach with their advantages and disadvantages.

Following this study, we can state that PF approaches provide a good quality phase image compared to MN approaches because operation using MN is only a general estimation. However, the PF approaches are slow and take more time.

Comparing the BC methods (with variants) and QG methods, we can state that the BC methods are faster than the QG methods but their major drawback is the isolated cluster regions without unwrapping, which is not the case for QG methods where all the regions are wrapped depending on local quality. Nevertheless, hybridization between the different methods is always possible in order to combine the advantages and finish with a compromise.

11

Compressed Sensing in Radars

11.1. Introduction

The CDMA (code division multiple access) multiplexing technique is the basis for "IS-95-B" norms or "CDMA One" for the second generation and "cdma2000" for the third generation, which are used in North America. It is also used in the "UMTS" (Universal Mobile Telecommunications System) norm, also called "W-CDMA" (wide-band CDMA), from the third generation of European mobile telephony. In addition, it is used in parts of the protocol in fourth-generation systems. For example, orthogonal sequences are used for multiplexing demodulation signals in the LTE's uplink. The DS/CDMA technique is expected to be used in fifth-generation systems at least in combination with other multiplexing techniques [YAN 16].

In a DS/CDMA system, different users are identified by different sequences of spread spectrums. The latter are not perfectly orthogonal, as they trigger interferences to multiple accesses (IAM) [SHU 15]. Because these systems offer high transmission rates, the IAM become remarkable. Moreover, these systems suffer from a vanishing signal caused by the multiple paths. These two phenomena are the most significant handicaps in the CDMA system's specifications. If both these problems are not controlled correctly, they can lead to a serious degradation in detection quality [YAN 16].

To better exploit the advantages of a DS/CDMA system, the receiver should be perfectly synchronized with the transmitter. Synchronization is generally carried out in two stages: sensing and monitoring. The main unit in any sensing receiver is the decision-making mechanism, which should use a threshold. In classical systems, this threshold is adjusted, and then fixed depending on environmental conditions. The latter are not stable in mobile and wireless communications, so the fixed threshold becomes incapable of providing satisfactory performances. To solve this problem, sensing systems with adaptive threshold based on constant false alarm rate (CFAR) algorithms were introduced [HAC 12]. The receiver with a CFAR is one of the adaptive thresholding techniques that have really given an impetus to the evolution of radars. Several algorithms are then suggested to improve performances in adaptive code sensing in CDMA systems.

The ordered statistics adaptive processor (OSAP) detector was suggested in [KIM 98a]. Using a system with a search series strategy, this detector's noise story is estimated by the *k*th ordered sample. The same authors have suggested the adaptive acquisition processor (AAP) detector, considering multi-path channels with Rayleigh vanishing [KIM 97]. In this case, the noise's power is estimated by the sum of the uncensored cells, after censoring *k* high power cells. In [KIM 98b], the authors applied the same principle considering a hybrid sensing system to minimize the average sensing time. So that such a detector offers better performances, the number of censored cells should be greater than the number of multiple paths in the reference channel. To do this, an automatic censoring adaptive processor (ACAP) was introduced in [AIS 08] and has proven more robust than the two cited previously.

The variety of reception antenna has been used broadly to combat degradation of compressed sensing systems' performances. In [OH 02], the authors suggested applying the cell-averaging CFAR (CA-CFAR) detector to a more complicated sensing system with a hybrid structure, using the variety of antennae. The same hypotheses are considered in [HAC 12], but using the HAPAC (hybrid acquisition processor based on automatic multipath cancellation) process.

Recently, a new adaptive sensing technique was suggested to improve performances in the series system [DOU 16]. This technique used threshold optimization, which is based on particle swarm optimization (PSO). All these techniques succeeded in improving performances in adaptive sensing, either by using more complex architectures, or integrating CFAR algorithms that require a fairly long processing time.

To minimize the average sensing time, double-dwell architecture was suggested in [OH 05], then combined in [KRO 08] with the hybrid search strategy. Both these systems use the CA-CFAR algorithm, which seriously affects the system's performances in non-homogenous environments. In [BEN 12], the authors suggested a CFAR algorithm based on artificial neural networks, and specifically multilayer perceptron (MLP), called artificial neural network CFAR (ANN-CFAR). These performances are evaluated in a non-homogenous additive white Gaussian noise (AWGN) channel, using a series sensing system. The same algorithm is applied in [BEN 13a] considering a multi-path channel with Rayleigh vanishing. This detector's performances are compared with CA-CFAR and order statistics CFAR (OS-CFAR) detectors in different environmental conditions. The results obtained show that this detector is more effective than CA-CFAR and OS-CFAR detectors, from the point of view of the probability of detection and average sensing time. With its fairly short processing time, this detector manages to improve the probability of detection but the sensing time remains fairly long, due to the search series strategy.

In this book, the ANN-CFAR algorithm based on multilayer neural networks is combined with the double-dwell search strategy. This is chosen because of: (1) its simple structure, (2) its minimal sensing time and (3) its reduced false alarm rate. In addition to improving the detection probability offered by use of the ANN-CFAR detector, the suggested system will be capable of considerably minimizing the mean sensing time due to the use of the double-dwell schema and the neuron detector's very short processing time.

11.2. Description of the system

The adaptive sensing system suggested in this book is formed of two adaptive detectors (ADs), linked in series, which have identical structures, as Figure 10.1 shows. Each AD contains two; the first block is a conventional non-coherent detector with a matched filter (MF). It is formed of two branches, and each contains an MF correlator. The value of the power's signal at the latter's output is high at the square, then added with the value of the other branch, hence the term, "non-coherent". The signal received passes through the non-coherent detector for the demodulation (here binary phase shift keying [BPSK]). It is then despread in the correlator, which is created based on an MF. The second block illustrates the decision process based on the CFAR principle, which uses the artificial multi-layer networks.

The two ADs are formed as follows: the first AD has a partial correlation of short duration compared to the second so the sensing system can reject the cells that are not in fast phase ($N_1 < N_2$, or and N_2 are the partial correlations of the first and second ADs, respectively). If the duration of the partial correlation is short, then the probability of a false alarm will be high. To compensate for this problem, the second AD has longer partial correlation duration. Consequently, if the decision process for the two ADs is controlled properly, a "double-dwell" system can, on the one hand, minimize the average sensing time and, on the other hand, improve the probability of detection while maintaining the probability of a false alarm at a desired minimal value.

For each AD, the non-coherent detector's output is sent in series to a shift register of length $M + 1$. The first register, represented by Y_i, ($i = 1, 2$), registers the result of multiplying the power of the input signal with the value of the partial correlation between the PN sequence entering and that generated locally. The following M registers, written Y_{ij}, $i = 1, 2, j = 1, 2, \ldots, M$, register the power of the previous M phases; they are called "reference windows".

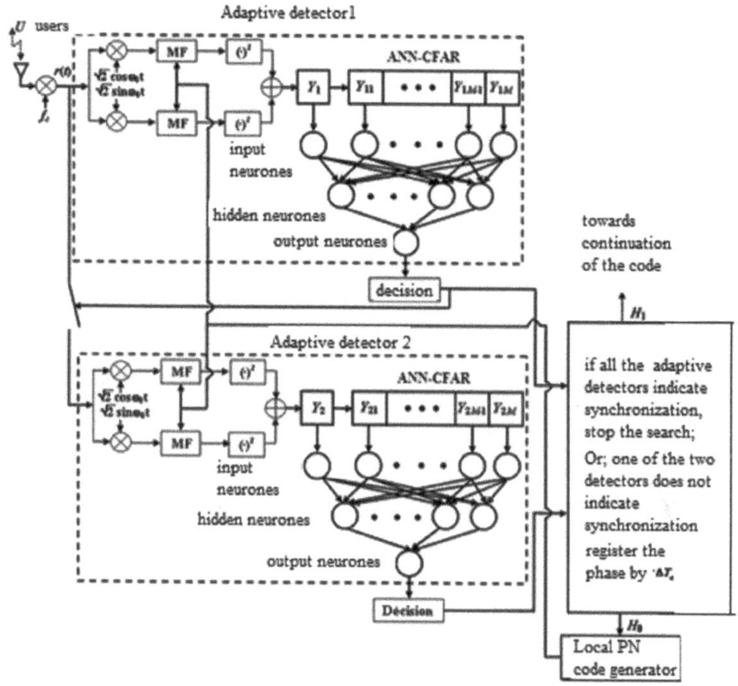

Figure 11.1. *Block diagram of the suggested system*

The system operates as follows:

The PN signal received, plus noise, interferences with multiple access and signals from multiple paths, arrives at the entrance to each AD:

– If the first AD indicates that the current cell is in phase, then the second AD begins to run. If the latter also indicates the synchronization, the monitoring loop is activated, and the phase of the local PN code is delayed by ΔT_C, where T_C is the duration of the chip from the PN sequence. In this case, the next cell is then examined and so on. Consequently, the distributions of the outputs from the two ADs are independent. The value of Δ is usually taken as 1/8, 1/4, 1/2 or 1; in this book Δ is equal to 1.

– In the reverse scenario, if at least one of the two ADs declares H_0 (non-synchronization), the phases of both PN codes (received and local) are automatically adjusted to the next position and the entire test is repeated.

11.3. System analysis

The communication system to be studied is formed of U users who can transmit simultaneously; it includes the $(U-1)$ users who transmit data (who have finished the sensing phase) and a single initial synchronization user (including the PN sequence acquired at the base station).

We assume that the first user is the initial synchronization user, for whom we wish to evaluate these performances. Each user is assigned a unique PN sequence, which spreads the sequence of data. $\left\{c_j^{(k)}\right\}$ indicates the binary sequences of $\{+1, -1\}$, where $c_j^{(k)}$ can take the values $+1$ and -1 with equal probabilities, and $c_k(t)$ denotes the shape of the signal of the PN sequence of the kth user, given by:

$$c_k(t) = \sum_{j=-\infty}^{\infty} c_j^{(k)} P_{T_c}(t - jT_c) \quad [11.1]$$

with:

$$P_\tau(t) = \begin{cases} 1, & 0 \leq t \leq \tau \\ 0, & if\ not \end{cases} \quad [11.2]$$

11.3.1. *Transmitter*

By using data sequences to modulate the shape of the PN sequence and the carrier, the signal transmitted from the kth user is expressed by [SOF 16]:

$$s_k(t) = \sqrt{2P_k}\, b_k(t) c_k(t) \cos(2\pi f_c t + \phi_k) \quad [11.3]$$

where, P_k represents the power of the signal transmitted from the kth user, $b_k(t)$ is the data sequence of the kth user, f_c is the clean pulse, whereas the phase φ_k is a random variable uniformly distributed along the interval $[0, 2\pi]$.

The presence of data modulation in the initial synchronization of the signal complicates the synchronization process for the code at the receiver. To do this, in many DS/CDMA systems, the transmitter facilitates the initial synchronization by transmitting the phase of the coded signal without data at the start of each transmission. In our analysis and to simplify the problem, we assume that there is no data modulation for the signals transmitted in the sensing phase.

11.3.2. *The receiver*

The signal received at the base station can be considered as the sum: (1) of the signal's different paths with initial synchronization, (2) of the data transmission signals (multiple access interference) and (3) of the additive white Gaussian noise; it can be expressed by the following expression [BER 14]:

$$r(t) = \sum_{l=0}^{L-1} \sqrt{2P_R}\,\alpha_{1l} c_1(t - \tau_1 - lT_c)\cos(2\pi f_c t + \theta_{1l})$$

$$+ \sum_{k=2}^{U}\sum_{l=0}^{L-1} \sqrt{2P_I}\,\alpha_{kl} b_k(t - \tau_k - lT_c) c_k(t - \tau_k - lT_c)\cos(2\pi f_c t + \theta_{kl}) + n(t) \qquad [11.4]$$

where $\{\tau_k\}$ are the relative delays associated with an asynchronous transmission scheme, and are modeled by random variables uniformly distributed along the interval $[0, T_c]$, while $n(t)$ is an additive white Gaussian noise with zero mean and power spectral density $N_0/2$. As the $(U - 1)$ users engaged are in the process of transmitting data, we assume that their signals have an ideally commanded power and that the mean power received from each engaged signal is P_I. The mean power received at the base station from the initial synchronization user is usually different to that from users who are in the data transmission phase; this power is expressed by P_R.

11.4. Transmission channel

In equation [11.4], we assume that the L weights $\{\alpha_{kl}\}$ are independent random variables and identically distributed (*i.i.d.*) with a Rayleigh probability density function (PDF). This is the most widely used model for a channel with vanishing caused by multiple paths, considering selective frequency. This PDF is given by [LIA 01]:

$$f_{\alpha_{kl}}(x) = \frac{x}{\sigma^2}\exp\left(\frac{x^2}{2\sigma^2}\right), \qquad x \geq 0 \qquad [11.5]$$

with $\alpha^2 = E[\alpha_{kl}^2] = 2\sigma^2$, where $E[\cdot]$ designates the expected value, and the phases $\{\theta_{kl}\}$ are assumed to be independent random variables and uniformly distributed over $[0, 2\pi]$ and they are independent of $\{\alpha_{kl}\}$. In addition, the integration time in the sensing process is usually shorter than the duration of one bit of information, that is $\tau_D \leq T$, so we suppose that the vanishing is slow enough to guarantee that the amplitudes of the chips during the integration time have undergone identical distortions. It is generally supposed, too, that the total power of the vanishing in all the multiple paths is standardized to the unit. Therefore, the mean power of the vanishing in each path is uniform or decreases exponentially with the rate ν; it can be expressed by [SHI 01]:

$$E[\alpha_{kl}^2] = \begin{cases} 1/L & , \quad \nu = 0 \\ \dfrac{1-\exp(-\nu)}{1-\exp(-\nu L)}\exp[-(l-1)\nu], & \nu \neq 0, \quad l = 1,2,\ldots,L \end{cases} \qquad [11.6]$$

11.5. Decision variables

According to Figure 11.1, the output of the mean filter of the in-phase branch (Y_I) can be expressed by:

$$\begin{aligned} Y_I &= \int_0^{NT_c} r(t)c_1(t-iT_c)\sqrt{2}\cos(2\pi f_c t)dt \\ &= \sqrt{P_R}NT_c\left[\sum_{l=0}^{L-1} S_I(l) + \sum_{k=0}^{U}\sum_{l=0}^{L-1} M_I(k,l) + \eta_I\right] \end{aligned} \qquad [11.7]$$

where $S_I(l)$ is the user of interest component, which is the sum of the component of the signal and the noise, and can be expressed as follows:

$$S_I(l) = \frac{\alpha_{1l} \cos\theta_{1l}}{NT_c}[\tau R_N(i, M+1) + (T_c - \tau)R_N(i, N)] \qquad [11.8]$$

We have $\tau_1 - (l-i)T_c = MT_c + \tau$, where τ_1 is the delay created by the first user, and it can be considered as a random variable uniformly distributed over [0, T_C]. In this equation, R_N (I, M) represents the partial auto-correlation of the spreading sequences, which is defined by:

$$R_N(i, M) = \sum_{\upsilon=0}^{N-1} c_{\upsilon+i}^{(1)}.c_{\upsilon+i+M}^{(1)} \qquad [11.9]$$

A non-coherent sensing system can acquire only one path of a multi-path signal at any given moment the signal's other paths (L – 1) are then interferences of the in-phase branch. Moreover, if the locally generated sequence is not in phase with any of these L paths of the desired signal all the L paths of the signal from the user of interest from interferences.

Random data modulation does not change the statistics of the random sequences because the random processes are assigned for all the system's users. Consequently, supposing $\tau_k + (l-i)T_c = NT_c + \tau$, where τ is a random variable uniformly distributed along the interval [0, T_C], the term of the multiple access interference, $M_I(k, l)$ can be expressed as follows:

$$M_I(k,l) = \frac{\sqrt{\rho}\alpha_{kl} \cos\theta_{kl}}{NT_c}[\tau R_N^{(k)}(i, M+1) + (T_c - \tau)R_N^{(k)}(i, M)] \qquad [11.10]$$

where $\rho = P_I / P_R$ and $R_M^{(k)}(i, M)$ represents the partial inter-correlation between the sequence of the kth user and the locally generated sequence, which is defined by:

$$R_N^{(k)}(i,M)=\sum_{\upsilon=0}^{N-1} c_{\upsilon+i}^{(1)}\cdot c_{\upsilon+i+M}^{(k)} \qquad [11.11]$$

Finally, the term of the phase noise due to the presence of additive white Gaussian noise $n(t)$ can be expressed by:

$$\eta_l=\frac{1}{\sqrt{P_R}NT_c}\int_0^{NT_c} n(t)c_1(t-iT_c)\sqrt{2}\cos(2\pi f_c t)dt \qquad [11.12]$$

The output from the correlator with mean filter of the branch in quadrature Y_Q can be obtained in the same way. As the system's spreading sequences are modeled by random binary sequences with the value ± 1 with equal probabilities, we can show that the limits of Y_I and Y_Q are asymptotically Gaussian, when the number of users increases (due to the central limit theorem). In our analysis, we use the Gaussian approximation and we model the internal thermal noise resulting from dispersion by multiple paths from the user of interest, the multiple access interference from the $(U-1)$ data transmission users and the thermal noise by an additive white Gaussian noise. Therefore, the mean value of component $S_I(l)$ is given by:

$$E[S_I]=\frac{3}{4}\alpha_{1l}\cos\theta_{1l} \qquad [11.13]$$

$S_I(l)$ is a random Gaussian variable with zero mean and a variance expressed by:

$$Var[S_I(l)]=\frac{\alpha^2}{3N} \qquad [11.14]$$

The term multiple access resulting from the kth user engaged can be approximated by a random Gaussian variable with zero mean and a variance given by:

$$Var[M_I(k,l)]=\frac{\rho\alpha^2}{3N} \qquad [11.15]$$

The term noise is a random Gaussian variable with zero mean and a variance expressed by:

$$Var[\eta_I] = \frac{1}{2N\overline{\gamma}_c} \quad [11.16]$$

where $\overline{\gamma}_c = P_R T_c / N_0$ represents the signal-to-noise ratio per chip (SNR/Chip).

So according to hypothesis H_1, the variance of Y_I can be expressed by [LIA 01]:

$$Var[Y_I] = \frac{(L-1)\alpha^2}{3N} + \frac{(U-1)L\rho\alpha^2}{3N} + \frac{1}{2N\overline{\gamma}_c} \quad [11.17]$$

If we suppose that hypothesis H_0 is evaluated, all the L access routes for the user of interest constitute interference. Consequently, the variance of Y_I can be calculated by the previous equation, if we substitute $(L-1)$ for L. However, as our aim is to analyze sensing in multiple access transmission environments, we suppose that the variance of Y_I under hypothesis H_0 is identical to that under hypothesis H_1.

The statistics of the component in phase quadrature Y_Q can also be obtained in the same way. Therefore, the normalized variance of Y_Q is $Var[Y_Q] = Var[Y_I]$.

The statistics of the correlators' outputs, Y_I and Y_Q for the branches in phase and in phase quadrature respectively, are approximately Gaussian random processes. Consequently, for conventional detection, the decision variable $Y_x = Y_I^2 + Y_Q^2$ represents an H_1 (synchro) state or an H_0 (non-synchro) state. Because of the Gaussian nature of Y_I and Y_Q, and supposing that Y_x forms a sample H_1, the PDF of Y_x is a chi-square distribution with two degrees of freedom, given by [OH 05, LIA 01]:

$$f_{Y_x}^{kl}(y_x \mid H_1) = \frac{1}{2\sigma_0^2} \exp\left(-\frac{m^2 + y_x}{2\sigma_0^2} \right) I_0\left(\frac{m\sqrt{y_x}}{\sigma_0^2} \right), \quad y_x \geq 0 \qquad [11.18]$$

where

$$\sigma_0^2 = Var[Y_I] = Var[Y_Q] = \frac{(L-1)\alpha^2}{3N} + \frac{(U-1)L\rho\alpha^2}{3N} + \frac{1}{2N\overline{\gamma}_c} \qquad [11.19]$$

$I_0(.)$ is the first-degree, order-zero, "modified Bessel function", and m^2 is the normal non-centrality parameters; it is given by:

$$m^2 = \frac{9}{16}\alpha_{1l}^2 \qquad [11.20]$$

When Y_x forms a sample H_0, its PDF can be expressed by:

$$f_{Y_x}^{kl}(y_x \mid H_0) = \frac{1}{2\sigma_0^2} \exp\left(-\frac{y_x}{2\sigma_0^2} \right), \quad y_x \geq 0 \qquad [11.21]$$

If we normalize the non-coherent detector's output $(Y_i = Y_x / \sigma_0^2)$, the PDF of Y_i, $i = 1, 2$, can be expressed by:

$$f_{Y_i}^{kl}(y_i \mid H_1) = \frac{1}{2} \exp\left(-\frac{S^2 + y_i}{2} \right) I_0\left(S\sqrt{y_i} \right), \quad y_i \geq 0 \qquad [11.22]$$

where $S = \sqrt{m^2/\sigma_0^2} = S_0.(\alpha_{1l}/\alpha)$, with S_0 given by [BEN 13b]:

$$S_0 = \frac{3}{4}\left(\frac{(L-1)}{3N} + \frac{(U-1)L\rho}{3N} + \frac{1}{2N\,\alpha^2\,\overline{\gamma}_c} \right)^{-1/2} \qquad [11.23]$$

The PDF of Y_i under hypothesis H_0 becomes:

$$f_{Yi}^{kl}(y_i \mid H_0) = \frac{1}{2} \exp\left(-\frac{y_i}{2} \right), \quad y_i \geq 0 \qquad [11.24]$$

After absorption of the random variable S, $f_{Y_i}^{kl}(y_i \mid H_1)$ becomes [DOU 16]:

$$f_Y^{kl}(y \mid H_1) = \frac{1}{2\left(1+\mu E[\alpha_{kl}^2]\right)} \exp\left(-\frac{y}{2(1+\mu E[\alpha_{kl}^2])}\right), \quad y \geq 0 \qquad [11.25]$$

where $\mu = 9\sigma^2 / 32\sigma_0^2$ [15], and:

$$f_{Y_i}^{kl}(y_i \mid H_0) = \frac{1}{2}\exp\left(-\frac{y_i}{2}\right), \quad y_i \geq 0 \qquad [11.26]$$

11.6. ANN-CFAR detector

The ANN-CFAR detector, shown in Figure 11.1, is formed of three layers: the input layer, the hidden layer and the output layer. The input layer carries (M + 1) neurons receiving, respectively, the (M + 1) variables $[y_i \; y_{i1} \; y_{i2} \; \cdots \; y_{iM}]$ given that variable y_i receives the power from the phase to be tested and the input number (M + 1) depends on the nature of the problem. If the number of input cells is relatively small (lower than 30), the performances of the ANN-CFAR detector degrade [CHE 06]. After several tests, it has been found that M = 32 provides good performances. The number of neurons in the hidden layer is chosen so as to optimize the learning time, and avoid the memorization effect, which is the consequence of a large number of hidden units. However, the drops in the number of neurons in the hidden layer lower the network's aptitude for learning. Therefore, the number of neurons in the hidden layer H is chosen equal to 5, with sigmoid transfer functions given by:

$$f(x) = \frac{1}{1+\exp(-x)} \qquad [11.27]$$

The output layer contains a single neuron with a sigmoid activation function. The detector's output is this neuron's output, it only takes two values: 1 (synchro) or 0 (non-synchro). The output neuron's biases are considered to be the detection threshold, which should be chosen in such a way as to fix the false alarm rate at a minimal desired value after the learning phase. The two systems' threshold values (series and double-dwell) for different values of the false alarm rate are summarized in Table 11.1.

P_{fa}	0.01	0.001	0.0001
Series system	6.13	6.8545	6.8623
Double-dwell system	1.5815	3.92	6.13

Table 11.1. *Values of θ of two systems for different values of Pfa*

The weight of connections between the *i*th (I = 1, 2, …, M + 1) neuron of the input layer and the *j*th (j = 1, 2, …, H) neuron in the hidden layer, and the weight of connections between the *j*th neuron of the hidden layer and the neuron of the output layer are written as w_{ij}^h and w_{j1}^o, respectively. The output of the *j*th hidden neuron is given by [CHE 06]:

$$h_j(t) = f_j\left(\sum_{i=1}^{M+1} w_{ij}^h x_i(t)\right) \qquad [11.28]$$

with f_j representing the activation function of the *j*th hidden neuron and $x_i(t)$ the input vector

$$u(t) = \sum_{j=1}^{H} w_{j1}^o h_j(t) \qquad [11.29]$$

$u(t)$ is the input of the output layer. The output from the detector ANN-CFAR is therefore the output of the neuron from the output layer $o(t)$, and is calculated by the following equation:

$$o(t) = \frac{1}{1+\exp\{-\alpha[u(t)-\theta]\}} \qquad [11.30]$$

where α is a real number of very low value and θ represents the detection threshold that enables us to adjust and fix the false alarm rate.

11.6.1. *Learning phase*

The back propagation algorithm is one of the most used supervised learning algorithms for PMC networks. This algorithm was introduced in the 1980s by Rumelhart [RUM 86]. Its principle is based on a modification of the synaptic weights from a back propagation of the error from the output to the entry layer, passing through the hidden layers. There is a set of 156 examples (learning base), formed of pairs (input, desired exit) following centered and non-centered chi-squared distributions for different parameters. At each stage, an example is shown at the entrance to the network. An output is then calculated and the calculation is carried out from the input layer to the output layer passing via the hidden layer; this procedure is called "forward propagation". The error is then calculated (quadratic sum of errors over each cell of output) and is back-propagated through the network giving rise to a change in weight. This process is repeated by successively presenting each example. If for all the examples the error is less than a chosen threshold, it is then said that the network has converged.

Learning consists of minimizing the quadratic error (considered as a function of weight) incurred over the set of examples by adjusting the weight. To do this, it is vital to consider the differentiable activation functions.

11.6.2. *Learning algorithm*

1) Weight initialization w_{ij}^h and w_{j1}^o of the hidden layer and the output layer, respectively, by the random values between 0 and 1, as well as the biases.

2) Introduce the input vector $X_p = [x_1, x_2, \ldots, x_{M+1}]^T$ to the neurons of the input layer.

3) Calculate the desired exit using equations [11.27]–[11.30].

4) Calculate the terms of error for the neurons of the output layer:

$$\delta_{pk}^{o} = (y_{pk} - o_{pk}) f_k^{o'}(net_{pk}^{o}) \qquad [11.31]$$

5) Calculate the terms of error for the neurons of the hidden layer:

$$\delta_{pj}^{h} = f_j^{h'}(net_{pj}^{h}) \sum_k \delta_{pk}^{o} w_{kj}^{o} \qquad [11.32]$$

6) Adjust the weight of the neuron of the output layer:

$$w_{kj}^{o}(t+1) = w_{kj}^{o}(t) + \mu \delta_{pk}^{o} i_{pj} \qquad [11.33]$$

7) Adjust the weight of the neurons of the hidden layer:

$$w_{ji}^{h}(t+1) = w_{ji}^{h}(t) + \mu \delta_{pj}^{h} x_{pi} \qquad [11.34]$$

8) Calculate the error:

$$E_p = \frac{1}{2} \sum_{k=1}^{m} \delta_{pk}^{2} \qquad [11.35]$$

9) Each time, we present the network with an input vector in a random fashion with their associated output and repeat the calculation process from stage 3.

10) Once we present the network with all the examples of the learning base, we calculate the following cost function:

$$E = \sum_{p=1}^{P} E_p \qquad [11.36]$$

If we reach the desired error, the learning process will terminate. If not, we start from stage 2 with all the examples of the learning base.

11.6.3. *Validation phase*

After the success of learning and setting the detection threshold, the network should be tested by data that are not in the learning base. The results obtained are then compared with cells from the CA-CFAR detector, by considering a homogenous medium. If this detector's performances are judged satisfactory, then it can be considered ready to be used as a CFAR detector. If not, the entire procedure should be repeated several times, changing the number of neurons in the hidden layer.

11.7. Probability of detection and of false alarm

Because analytical methods are not available to evaluate the ANN-CFAR detector's performances, the probabilities of a false alarm and detection for the two ADs have been calculated by simulation, using the MATLAB software. The positions of the cells to be tested are normally known during the simulations. P_D and P_{FA} are calculated by the following two equations, using the Monte-Carlo method.

$$P_{FA_i} = \frac{Number\ of\ false\ detections}{total\ number\ of\ noisy\ cells\ tested},\ \mathrm{i} = 1, 2 \qquad [11.37]$$

$$P_{D_i} = \frac{Number\ of\ detections}{total\ number\ of\ cells\ tested} \qquad [11.38]$$

As the outputs of the two ADs' correlators are independent, the system's probability of detection is the product of the two ADs' probabilities of detection [OH 05]. Consequently, the probability of detecting the correct cell is given by:

$$P_D = P_{D_1} \cdot P_{D_2} \qquad [11.39]$$

For the same reasons, the probability of a false alarm can be expressed by:

$$P_{FA} = P_{FA_1} . P_{FA_2} \quad [11.40]$$

11.8. Mean acquisition time

To derive the expression of the mean acquisition time, we take account of the following hypotheses:

1) There is a single sample corresponding to the correct phase (a single cell H_1).

2) All the samples are independent.

3) We choose the lengths of the partial correlations N, N_1 and $N_2 >> 1$ so they give a correlation value (between the PN sequence received and that generated locally) around zero when the two sequences are not in phase (cell H_0).

4) The region of uncertainty is equal to the length q of the PN code.

The mean acquisition time can be induced as [OH 05], using the following equation:

$$\overline{T_{acq}} = \frac{N_1 T_c}{P_D} + \frac{N_2 T_c}{P_{D2}} + (q-1)[N_1 T_c + N_2 T_c P_{FA1} + K(N_1 + N_2) T_c P_{FA}] \left(\frac{1-P_D}{P_D^2} + \frac{1}{2} \right) \quad [11.41]$$

where K is the penalty time associated with a false alarm.

11.9. Results and discussions

In this section, the suggested adaptive acquisition system's performances are evaluated with the help of IT simulations carried out using the Monte-Carlo technique, using 10^5 independent tests for each calculation, and by considering the following hypotheses: (1) the number of reference cells for each AD, M = 32, (2) the chip's duration, T_c = 1 µsec, (3) a periodic PN code of length q = 1,024 and therefore an uncertainty region of q cells, (4) a penalty time equal to KNT_c second, with a penalty constant K = 1,000 and (5) multiple paths with a uniform decay profile (ν = 0). The system's performances are analyzed for a slow vanishing channel that follows a Rayleigh distribution. Two performance criteria are considered: the probability of detection and the mean acquisition time. The results obtained are compared with those of adaptive serial system using the ANN-CFAR detector. To obtain reliable comparisons, the same environmental conditions are considered for both systems.

In Figure 11.2, we present the suggested system's probability of detection (double-dwell) as well as that of the serial system in a homogenous medium, that is in the presence of a single path and single user in the channel, with the false alarm rate as a performance parameter. As expected, we observe that if the false alarm rate is high, then the probability of detection is better. We also note that the suggested system is more effective than the serial system, because the mean gain obtained in relation to the system offered is around 3 dB, that is the suggested system offers the same probability of detection for a signal three times less powerful, which is important for detection.

To demonstrate the improvement brought about by the suggested system, we have made a comparative study with a serial system, relying on the criterion of the mean acquisition time. Figure 11.3 shows a set of curves of the average acquisition time depending on the two systems' (serial and double-dwell) SNR/Chip in a homogenous environment. It is clear that an increase in the false alarm rate increases the mean acquisition time. It should be noted that the results obtained clearly show that our system is very fast compared to the serial system. The suggested system is less than 1,000 times faster, which is very important for the acquisition system's performance. This is the reason why we have chosen the double-dwell system.

In the rest of the results, we have chosen to fix the false alarm rate at 10^{-3}, because this value represents the best compromise between the increase in the probability of detection and the decrease in the mean acquisition time, for both systems.

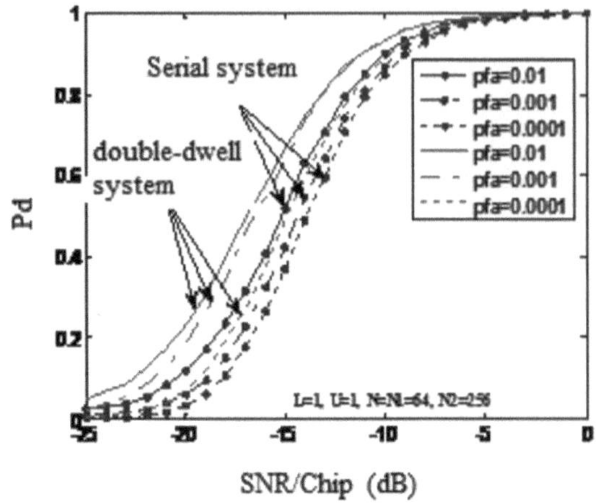

Figure 11.2. *Probability of detection depending on the two systems' SNR/Chip for different values of P_{fa}. For a color version of the figure, see www.iste.co.uk/benslama/networks.zip*

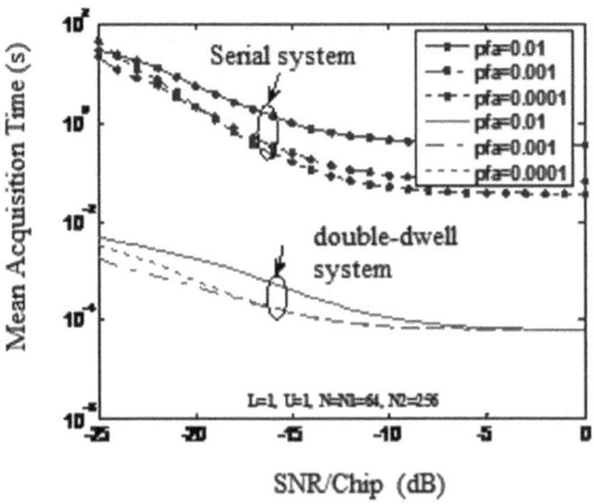

Figure 11.3. *Mean acquisition time depending on the two systems' SNR/Chip for different values of P_{fa}. For a color version of the figure, see www.iste.co.uk/benslama/networks.zip*

Figure 11.4 shows the two systems' probability of detection depending on the SNR/Chip, this characteristic is parameterized by the length of partial correlation (N, N_1 and N_2) in a homogenous medium. We note that the increase in this parameter increases the two systems' probability of detection. In addition, the suggested system gives better performances compared to the serial system for all the values considered of N, N_1 and N_2.

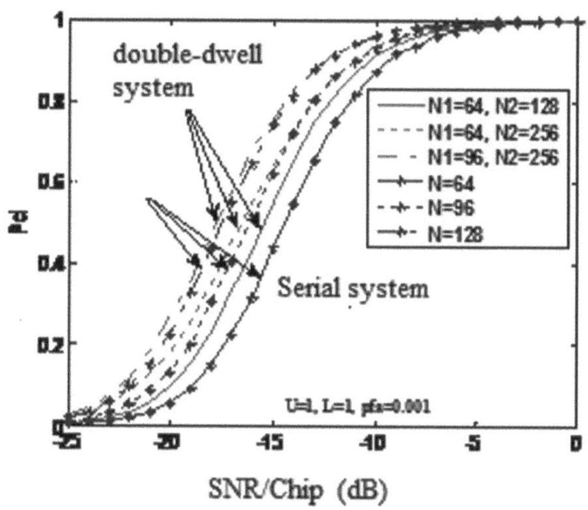

Figure 11.4. *Probability of detection depending on the two systems' SNR/Chip for different lengths of partial correlation. For a color version of the figure, see www.iste.co.uk/benslama/networks.zip*

In Figure 11.5, we present a comparative study of the two systems' average acquisition time considering N, N_1 and N_2 as parameters. We can clearly observe that, whatever the values of N, N_1 and N_2, the suggested system is more effective than the serial system in terms of rapidity in acquiring the PN code (at least 1,000 times faster). We also note that the increase in N increases the acquisition time of the serial system compared to the double-dwell system for higher values of SNR/Chip at –15 dB. Because the main objective of this work is to minimize the acquisition time, N will be equal to 64 in the results that follow.

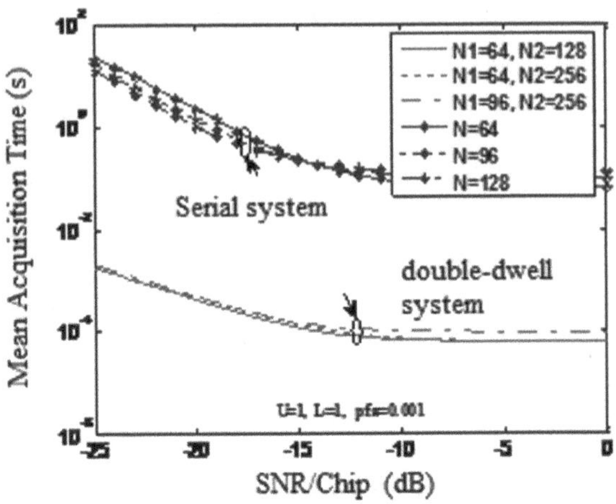

Figure 11.5. *Mean acquisition time depending on the two systems' SNR/Chip for different lengths of partial correlation*

From the same perspective, the effect of the variation in the lengths of the suggested system's integration sequence (N_1 and N_2), for the probability of detection and the acquisition time, is shown in Figures 11.6 and 11.7, respectively. We see that with the increase of N_2, the probability of detection improves but the acquisition time increases slightly when the SNR/Chip is lower than –10 dB, then stabilizes a minimal value for values of the SNR/Chip higher than –10 dB whatever N_2. On the other hand, the increase in N_1 increases the probability of detection, but it does not affect the acquisition time for weak values of SNR/Chip (< –15 dB). On the contrary, if the value of SNR/Chip exceeds –15 dB, the mean acquisition time increases with the increase in N_1. From these two figures, the suggested system's best performances are obtained for N_1 and N_2 equal to 64 and 256, respectively. Therefore, in the remainder of the results, these two values will be taken.

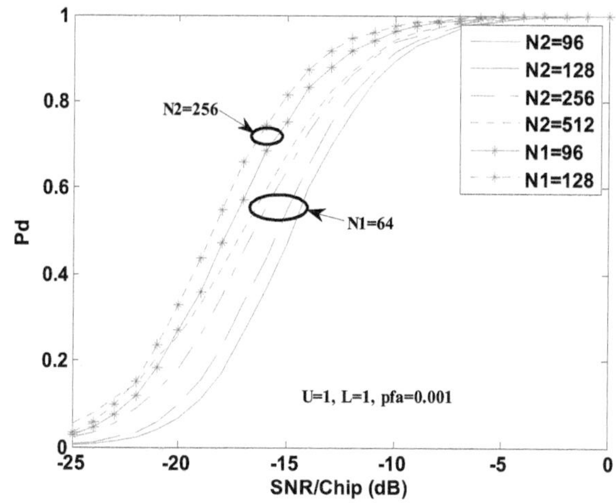

Figure 11.6. *Probability of detection depending on the suggested systems' SNR/Chip for different lengths of partial correlation*

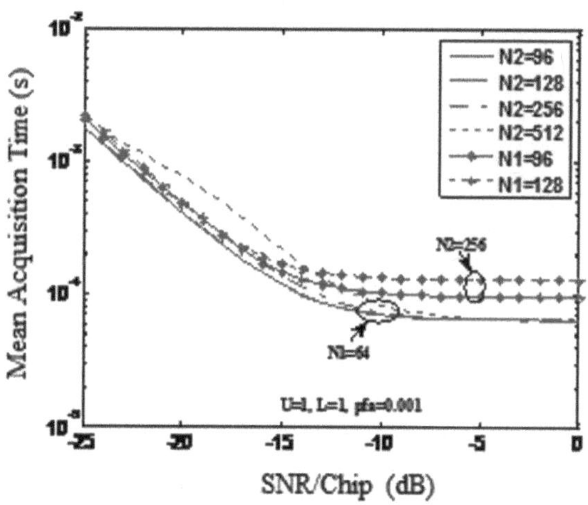

Figure 11.7. *Mean acquisition time depending on the suggested systems' SNR/Chip for different lengths of partial correlation*

In the second part of this chapter, we have tested the suggested systems' performances in a non-homogenous medium that is characterized by the presence of multiple paths and multiple users in the channel. Figure 11.8 shows the significant degradation imposed by the presence of multiple paths on the serial system's probability of detection compared to the double-dwell system, and this is for three cases of SNR/Chip (–8, –5, and 0 dB). This degradation increases with the increase in the number of paths. Therefore, if the number of paths increases, then the suggested system becomes more effective.

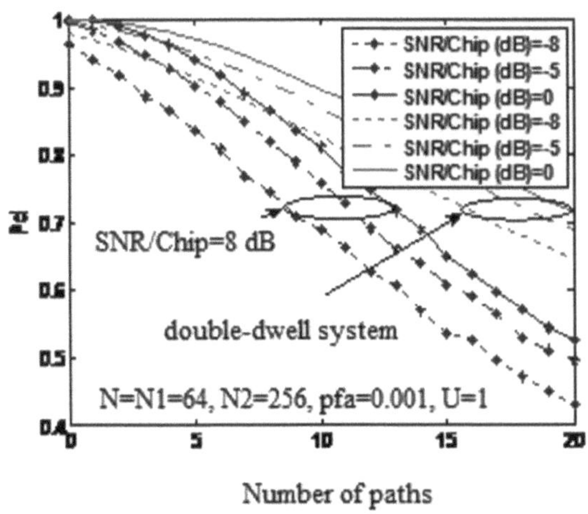

Figure 11.8. *Probability of detection depending on the number of the two systems' paths for different values of the SNR/Chip. For a color version of the figure, see www.iste.co.uk/benslama/networks.zip*

We now consider a comparison, in a non-homogenous medium, between the two systems' averages in acquisition time, depending on the number of multiple paths, for the three values of SNR/Chip cited previously (see Figure 11.9). As we can expect, the increase in the number of paths increases the two systems' mean acquisition time, but this increase is weaker for the suggested system. We also observe that a slight increase in the mean acquisition time compared to the serial system is caused by the decrease in the value of the SNR/Chip.

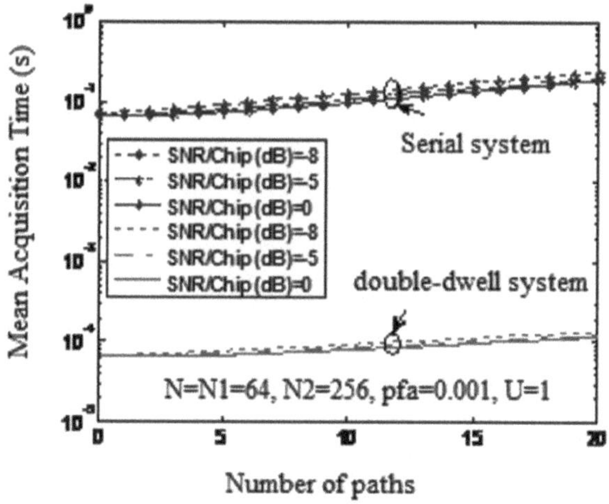

Figure 11.9. *Mean acquisition time depending on the number of the two systems' paths for different values of SNR/Chip. For a color version of the figure, see www.iste.co.uk/benslama/networks.zip*

To illustrate the effect of changing the number of users on the two systems' probability of detection, we consider three different values of SNR/Chip (–8, –5, and 0 dB). From the curves in Figure 11.10, we see that the double-dwell system is more effective than the serial system, because the degradation rate in the probability of detection is lowered with the increase in the number of users. Moreover, we note that the probability of detection degrades with the decrease in the value of the SNR/Chip, but in all situations, the suggested system's probability of detection remains the best, which was probably predictable.

Figure 11.11 shows the variation in mean acquisition time depending on the number of users for different values of SNR/Chip, considering both systems. This figure shows that the double-dwell system always gives good results compared to the serial system whatever the number of users and the value of the SNR/Chip.

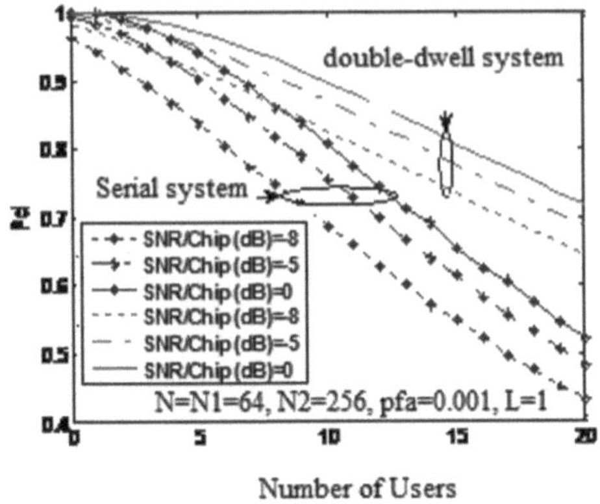

Figure 11.10. *Probability of detection depending on the two systems' number of users for different values of SNR/Chip. For a color version of the figure, see www.iste.co.uk/benslama/networks.zip*

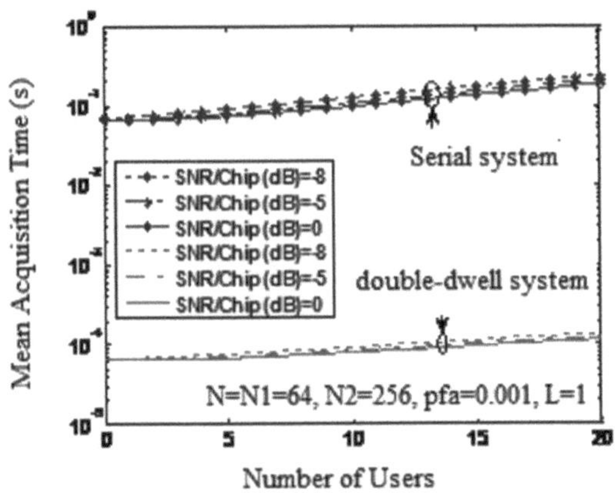

Figure 11.11. *Mean acquisition time depending on the number of users for two systems for different values of SNR/Chip. For a color version of the figure, see www.iste.co.uk/benslama/networks.zip*

We show in Figures 11.12 and 11.13 the influence of the variability of the ρ ratio (the power of the interferences [P_I] over the power of the useful signal [P_R]) from the point of view of the probability of detection and mean acquisition time, respectively. The results obtained show that the two systems' probability of detection decreases with the increase in the value of the ρ ratio, but the suggested system's rate of decrease is less than that of the serial system. It is clear that the suggested system's P_d is better than that of the serial system whatever the value of the ρ ratio. The variation in the ρ ratio has a weak influence on the suggested system's mean acquisition time; on the contrary, it slightly increases the serial system's mean acquisition time.

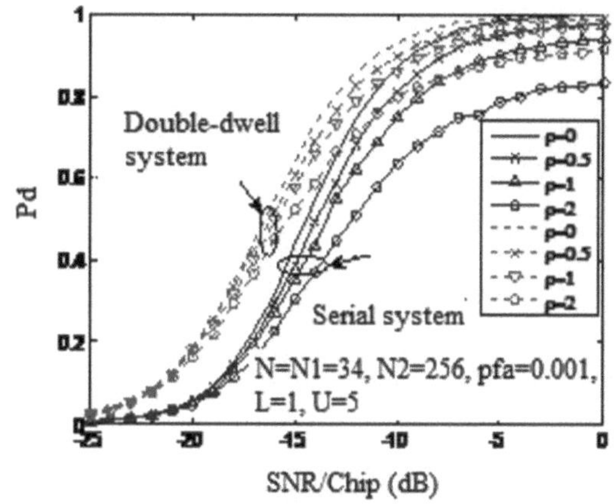

Figure 11.12. *Probability of detection depending on the two systems' SNR/Chip for different values of ρ. For a color version of the figure, see www.iste.co.uk/benslama/networks.zip*

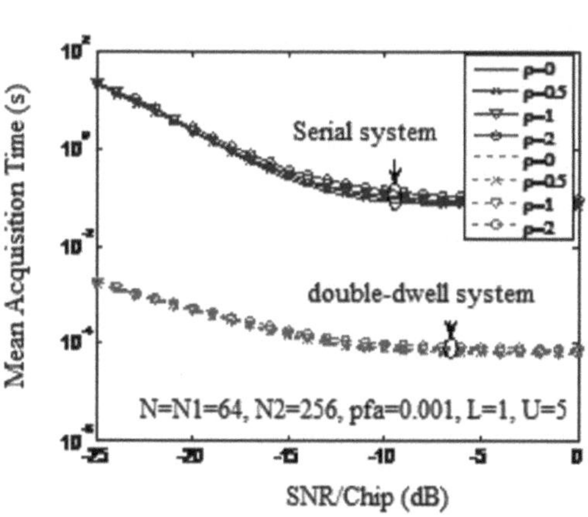

Figure 11.13. *Mean acquisition time depending on the two systems' SNR/Chip for different values of ρ. For a color version of the figure, see www.iste.co.uk/benslama/networks.zip*

11.10. Conclusion

In this book, we have discussed the problem of the adaptive acquisition of PN pseudo-random sequences used in DS/CDMA communication systems, where the communication requires synchronization between the transmitter and the receiver, which is carried out in two stages: acquisition and tracking. Our aim was to improve performances in the acquisition of PN codes (which is a more difficult task in the systems considered), in terms of the probability of detection and mean acquisition time. To reach this objective, we have suggested a system that combines the double-dwell architecture, characterized by its simple structure, its very short acquisition time and its minimal false alarm rate, and the ANN-CFAR detector that has given good performances with the serial architecture, and which is characterized by a fairly short processing time.

With the help of digital simulations, we have analyzed the performances of the suggested system depending on different parameters. We have considered the homogeneity (the absence of multiple access interferences caused by the multiple users and multiple paths) and the transmission channel's non-homogeneity (presence of multiple access interferences). Relying on the neuronal detector, the probability of a false alarm was kept fixed. Then, we compared its performances to a serial system, where we tried to minimize the mean acquisition time, without losing the improvement provided by the ANN-CFAR detector on the probability of detection, above all for weak values of SNR/Chip.

The results obtained clearly show that the suggested system is more effective and more robust than the serial system, more especially in the presence of multiple paths and multiple users, above all from the point of view of the mean acquisition time. It is more than a thousand times faster than the serial system.

12

Compressed Sensing in Electromagnetism

12.1. Introduction

Several problems resulting in electromagnetism can be formulated directly or conveniently positioned for an effective solution in the context of compressed sensing. This is motivated by the interest in the development and application of methodologies in compressed sensing for several conventional scenarios in electromagnetism. Massa *et al.* offer a review article [MAS 15] summarizing the state of the art of formulations in compressed sensing in electromagnetism. The diagnostic and synthesis of antenna networks, estimation of the beams' arrival direction, the solution of the inverse problems in electromagnetic diffraction and problems linked to radar imaging are covered.

12.2. Compressed sensing in electromagnetism

Compressed sensing techniques are built on the supposition that several physical quantities are intrinsically or extrinsically sparse; they can be represented by the developments of non-null coefficients.

In this chapter, we will show that it is possible to reformulate the problem of least squares minimization formulated by Davies [DAV 73] and Fuchs [FUC 09] in the case of planar microwave structures.

Compressed sensing is a new paradigm in data acquisition that enables an efficient and exact sparse representation of data using the techniques of sparse reconstruction based on nonlinear interpolation.

The certain advantages offered by compressed sensing in comparison with the approach by classic Shannon sampling, have ensured a fast diffusion of this theory in practically all branches of engineering and applied physics.

In [MAS 15], an important property has been discussed; this is the restricted isometry [CAN 08a]. The importance of this property is that it records the algebraic aspect with the geometric aspect. In fact, in compressed sensing or sparse reconstruction, an x vector is called S sparse if the number of its non-null elements does not exceed S. From a geometric point of view, a sparse vector S leads to a series obtained by joining all Poisson brackets $\binom{N}{S} = \frac{N!}{S!(N-S)!}$ of the S dimension subspaces in CN. This series will be called ΣS.

This aspect is discussed in [MIG 14, MIG 15] on sparse unknowns and their relationship with the restricted isometry property.

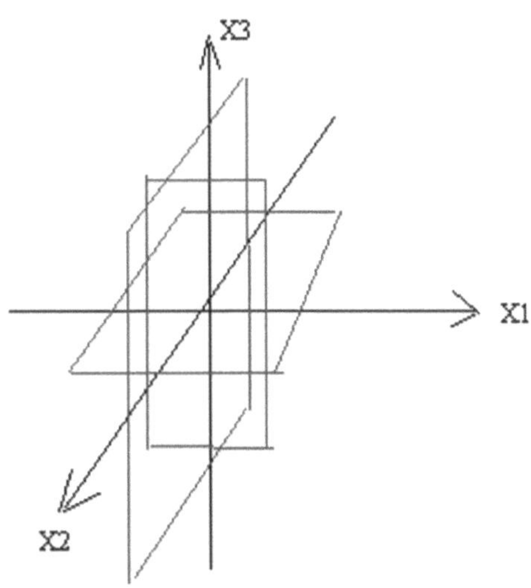

Figure 12.1. *The set Σ_2 of 2 sparse vectors (S = 2) of three elements (N = 3) is a union of the nonlinear subspaces of three two-dimensional subspaces*

For example, in Figure 12.1, the case of N = 3, S = 2 in the real space is known. This imaged view shows that ΣS takes a small part of the three-dimensional space as an essential consequence, and while a non-sparse vector x ∈ Σ_3 can be represented by any point in the 3D space, a bisparse (biparsimonious) vector can only be one of the three co-ordinates of the plan shown in Figure 12.2.

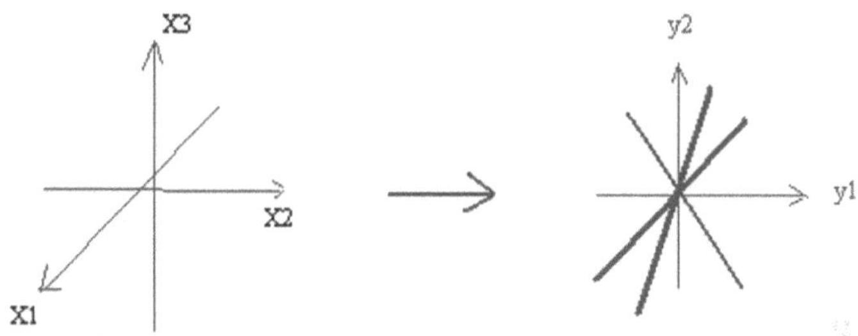

Figure 12.2. *Drawing of Σ_1 from three to two dimensional. Σ_1 is drawn in three lines; if the three lines are distinguable, then the drawing is invertible*

Oliveri *et al.* give interesting applications in electromagnetism. We note mainly the applications concerning detection of breakdowns in electromagnetic viability. The author confirms that the restricted isometry property cannot be checked in most electromagnetic problems [OLI 13].

Wang *et al.* provide an analysis of electromagnetic diffraction using basic functions combined with compressed sensing [WAN 13]. We compare the results obtained by the author with our work to gauge the improvements provided by compressed sensing [BEN 14].

It should also be noted that the theory of compressed sensing is used to reduce the quantity of data needed to reconstruct a network (RCS) in electromagnetic software [BUR 12].

To conclude this section, we will not omit the citation of the work of Cao *et al.* on the application of compressed sensing to solve three-dimensional problems in the electromagnetic diffusion of structures [CAO 12].

12.3. Application of compressed sensing and electromagnetism: a solution for diffusion using an arbitrarily formed conductive structure by the moment method

12.3.1. *Introduction*

The aim of this chapter is to tackle the electromagnetic diffusion of a conductive body of arbitrary shape in three dimensions illuminated by a flat wave. The analysis uses the electromagnetic field integral equation (MFIE), which is solved using the direct moment method with cellular modeling, in which the surface of the diffusion body is discretized into small cells with trilateral surfaces. The characteristic of the moment method used in formulating the problem is the point adaptation method.

12.3.2. *Matric formulation*

Considering the surface of the diffuser to be divided into small sections of triangular surfaces, the current in each patch is solved by the following equation:

$$J_s = 2\hat{n}\times H^i(r) + 2\hat{n}\times\oint_S J_s \times \nabla' g \, ds' \qquad r \in S \qquad [12.1]$$

where:

– $\oint$ = principal value of the surface's integral;

– H_i = flat wave of the incident field from a flat source.

The unknown Js is approximated by surface cell currents of the subsection:

$$J_s = \sum_{n=1}^{N} J_n(r) \qquad [12.2]$$

where

$$J_n(r) = I_n^1 P_n^1(r) + I_n^2 P_n^2(r) \qquad [12.3]$$

and

$$P_n^i(r) = \begin{cases} \hat{u}_n^i, \text{for } r \in \Delta S_n \\ 0, \ \text{elswhere} \end{cases} \qquad [12.4]$$

$$\text{for} \quad i = 1,2; \quad n = 1,2,\ldots N$$

where:

– ΔSn = surface of the *n*th patch of the N triangular patches;

– In1, In2 = unknown currents to be determined on ΔSn;

– $\hat{u}_n^1$, $\hat{u}_n^2$ = two arbitrarily orthonormal units chosen on the ΔSn plane.

A simple manner for defining unit vectors and to choose them in phase with the incident wave is such that

$$\hat{u}_n^1 = \frac{\hat{n}_n \times H^i(r_n)}{\left|\hat{n}_n \times H^i(r_n)\right|}, \qquad \hat{u}_n^2 = \hat{n}_n \times \hat{u}_n^1 \qquad [12.5]$$

where:

– r_n = vector position moving from the origin to the center of the patch ΔSn;

– $\hat{n}_n$ = normal unit vector at the surface ΔSn oriented toward the exterior of the closed surface S.

$$\hat{n}_n = \frac{(r_r - r_q) \times (r_p - r_r)}{\left|(r_r - r_q) \times (r_p - r_r)\right|} \qquad [12.6]$$

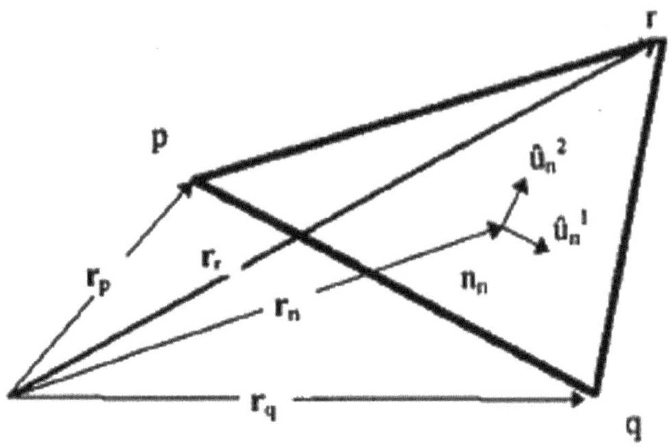

Figure 12.3. *Triangular patch ΔSn*

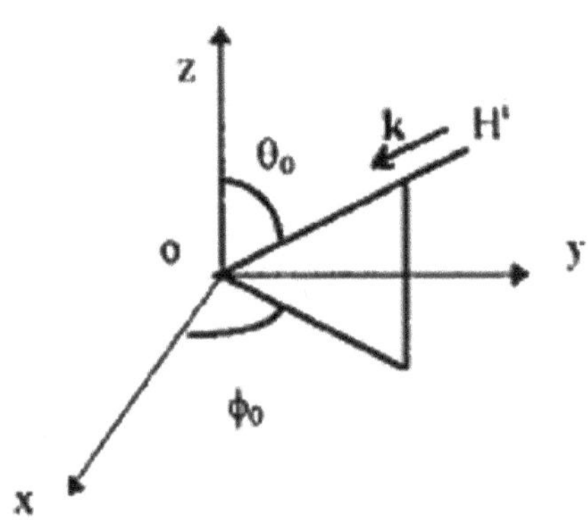

Figure 12.4. *Flat wave and its propagation vector*

Equation [12.1] is rewritten following the operator:

$$L\left[J_s(r)\right] = 2\hat{n} \times H^i(r) \qquad r \in S \qquad [12.7]$$

By substituting Js from equation [12.3] into the previous equation, we obtain:

$$\sum_{n=1}^{N}\sum_{j=1}^{2} I_n^j L\left[P_n^j(r)\right] = 2\hat{n}\times H^i(r) \qquad r\in S \qquad [12.8]$$

where

$$L\left[P_n^j(r)\right] = P_n^j(r) - 2\hat{n}\times\left[\hat{u}_n^j\times\int_{\Delta S}\nabla' g(r,r')\, ds'\right] \qquad r\in S \qquad [12.9]$$

Equation [12.7] has been approximated by equation [12.8], and for finite N, the left side of equation [12.8] is equal to ($2\hat{n}\times H^i(r)$) only in the direction of approximation. Our aim now is to choose the coefficients I^j_n, where j = 1, 2 such that equation [12.8] is met as closely as possible. There are 2N coefficients I^j_n; also, a system of 2N equations should be obtained, which makes it possible to determine these unknown constants. The procedure consists of equalizing both members of equation [12.8] for N different points of r on the surface of the diffuser (the centers of the patches). This point adaptation procedure is created via the following choice of weight functions $W_m{}^i$ as:

$$W_m^i = \delta(r - r_m)P_m^i(r) \quad i = 1,2\,; \qquad m = 1,2...N \qquad [12.10]$$

This also involves defining a symmetric product between P and Q:

$$\langle P,Q\rangle = \int_S P\cdot Q ds \qquad [12.11]$$

By taking the symmetric product with W_m^i in equation [11.8] for m = 1, 2…N and i = 1, 2, we obtain the following matrix equations:

$$\sum_{n=1}^{N}\sum_{j=1}^{2} I_n^j\langle W_m^i(r), L\left[P_n^j(r)\right]\rangle = \langle W_m^i(r), 2\hat{n}\times H^i(r)\rangle \qquad [12.12]$$

$$m = 1, 2\ldots N i = 1, 2$$

The series of previous equations can be written as:

$$\sum_{n=1}^{N}\sum_{j=1}^{2} Z_{mn}^{ij} I_n^j = V_m^i \qquad i=1,2 \qquad m=1,2...N \qquad [12.13]$$

The matrix elements are evaluated as follows:

$$V_m^i = \int_S \delta(r-r_m)\ P_m^i(r)\cdot 2\hat{n}\times H^i(r)\,ds \qquad [12.14]$$

$$Z_{mn}^{ij} = \int_S \delta(r-r_m)\ P_m^i(r)\cdot\left\{ P_n^j(r) - 2\hat{n}\times\left[\hat{u}_n^j \times \int_{\Delta S} \nabla' g(r,r')\,ds'\right]\right\} ds \qquad [12.15]$$

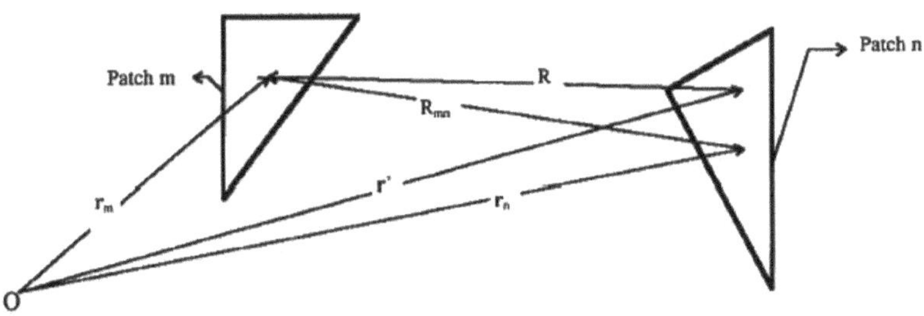

Figure 12.5. *Vectors linking patches m and n*

To summarize, the series of equations to be solved is:

$$\sum_{n=1}^{N}\sum_{j=1}^{2} Z_{mn}^{ij} I_n^j = V_m^i \qquad i=1,2 \qquad m=1,2...N \qquad [12.16]$$

$$V_m^i = \hat{u}_m^i \cdot 2\hat{n}_m \times H_i(r_m) \qquad [12.17]$$

$$Z_{mn}^{ij} = \delta_m^n \delta_i^j - \frac{1+jkR_{mn}}{2\pi R_{mn}^2} \exp(-jkR_{mn})\,\Delta S_n\left[\hat{u}_m^i \cdot \hat{n}_m \times \left(\hat{u}_n^i \times \hat{R}_{mn}\right)\right]\left(1-\delta_m^n\right) \qquad [12.18]$$

where δ_m^n is the Kronecker symbol equal to 1 if n = m and zero elsewhere.

12.3.3. *Convergence of the method of moments*

By applying the method of moments to convert an integral equation into a matrix equation, it is natural to ask how many of the test functions are needed in order to obtain precise enough results when the matrix's equation is solved. An often-used criterion is the following: if the solution remains reasonably constant in repeated calculations, in which the number of test functions is progressively increased, the result is considered to be acceptable. In the diffusion problem discussed in this chapter, the convergence of the MM (convergence of the radar current section) is a dependent segmentation structure; it is therefore studied as the number of correctives defining the diffusion body. The use of several sample currents over the body can lead to more precise RCS values not only because the current variation can be mapped with more precision, but because at the same time the digital integration is also carried precisely on the result of the decrease in the patch's size. Bearing in mind that when the frequency of the incident wave increases, the dispersion body becomes important in terms of wavelength; the main concern is to determine the upper limit frequency for which the representation of the surface element is enough.

The following merit factor can be used to ensure convergence [BEN 14]:

$$error(n) = \frac{\sqrt{\int_0^{2\pi} |\sigma_n(\theta) - \sigma_{nn-1}(\theta)|^2 d\theta}}{\int_0^{2\pi} \sigma_n(\theta) d\theta} \qquad [12.19]$$

where $\sigma_n(\theta)$ is the radar cross-section (RCS) and n – 1 and n represent the successive densities of the patch model.

12.3.4. *The solution's non-uniqueness*

A potential problem that may be encountered in the digital evaluation of diffusion properties of closed, solid conducting surfaces caused by an external source using the MFIE, is the excitation of an interior resonance that can lead to a numerical parasitic result for the dispersed field. This problem also arises in use of EFIE. This parasite phenomenon, which the digital result suddenly increases by hundreds of times above the correct value, is due to the non-uniqueness of the MFIE and EFIE. In the case of MFIE, the operator in equation [10.7] is singular; the tangential electric field on the conducting surface is not reduced to zero at or close to the interior problem's resonance frequencies. These integral equations' parasite resonances are eliminated by special techniques such as the extended boundary condition method, the combined field or source method (CFIE) or augmented integral equations (AEFIE, AMFIE) [TSA 02].

12.3.5. *Application on a sphere*

Our attention has focused on formulation of the diffusion problem using a three-dimensional conducting structure of arbitrary shape using a MM–MFIE solution. In this section, the method will be applied to a number of bodies of increasing complexity. These include some geometries that are generally the most demanding tests on digital technology. The geometry analyzed is a sphere with the radius a, as shown on Figure 12.6.

12.3.5.1. *Modeling the patch*

The patch model is first created by allocating points on the diffuser's surface. To obtain a good approximation of the diffuser's surface and to optimize the success of the digital technique, the following considerations can be made:

– the patches must be roughly equal, and in their region and centers as equidistant as possible;

– the facetted diffuser (which simulates the geometry) should also correspond to the study diffuser.

12.3.5.2. *Choosing parameters*

Each of the above-mentioned dispersion bodies represents a revolving body (around axis z). The dispersion body is illuminated by a linearly polarized incident plane wave. We will be interested in c-polarized 3D models of the bistatic radar cross-sections (θθ, ΦΦ). The radar section (back-scatter, bistatic and forward scatter RCS) models for two principal polarizations are designated "θθ" (E-plane) and "ΦΦ" (H-plane). The case in E-plane is produced when the two incident and dispersed fields are situated in the plane defined by the direction of propagation and the structure's axis of rotational symmetry. The case of the H-plane is defined similarly, with the magnetic field in place of the electrical field. Because of the axial symmetry, the crossed polarizations (θΦ and Φθ) are null and the copolarizations (θθ and ΦΦ) are independent of the azimuth angle.

12.3.5.3. *Checking validity*

The validity of digital results for RCS models for a diffusion structure calls for independent validity checks, which may be analytical, digital or experimental. Analytical solutions are available for only some simple shapes. Digital solutions obtained by independent means can also serve as validity checks and are evidently not limited in application to only some geometries. Perhaps the most convincing validity checks of all are a comparison with experimental measurement. There are also internal coherence checks that can be used to determine the likely degree of precision of the calculated results. The easiest to apply is that related to reciprocity. Reciprocity in a structure's bistatic diffusion motif or between motifs when the observation and incidence directions are interchanged, is required for solutions of Maxwell equations.

12.3.5.3.1. Modeling schema of the patch

Two patch distinct schemas are used for the sphere:

– 112 patches with 58 points;

– 120 patches with 62 points.

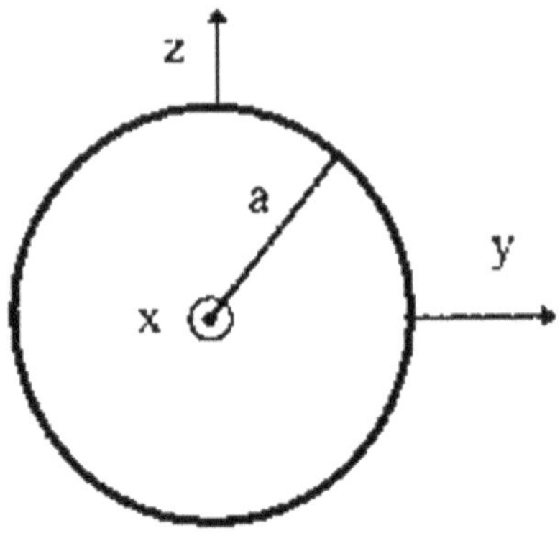

Figure 12.6. *Schema of a sphere with radius a (1 m)*

In the first schema, the sphere has been segmented in eight bands with angles equal to θ (polar angle) and each band, starting at the pole, is divided into patches as follows: 6, 12, 18, 24, 24, 18, 12, 6. In the second schema, the sphere has been segmented into 10 bands with increments equal to θ and each band starting at the pole is divided into patches as follows: 5, 10, 10, 15, 20, 20, 15, 10, 10, 5.

12.3.5.4. *Results and commentary*

A study of the convergence of the digital procedure will be shown in the figures.

Figures 12.7(a)–(d) show the SURFSCST calculations of bistatic, back-scatter and forward scatter RCS radar sections of a sphere with radius (a = 1 m) depending on the frequency of the incoming plane wave. The two lines on each figure are calculations for which the sphere has been divided into 112 and 120 patches corresponding, respectively, to 224 and 240 unknowns. The frequency range is from 10 to 300 MHz (corresponding to ka = 5.25, k = 2π / λ); there is not much difference between calculations of 112 and 120 patches and we can consider the latter to be convergent. A relative error $|\delta\sigma|$ that does not exceed 15% is thought to be a satisfactory criterion for relative convergence.

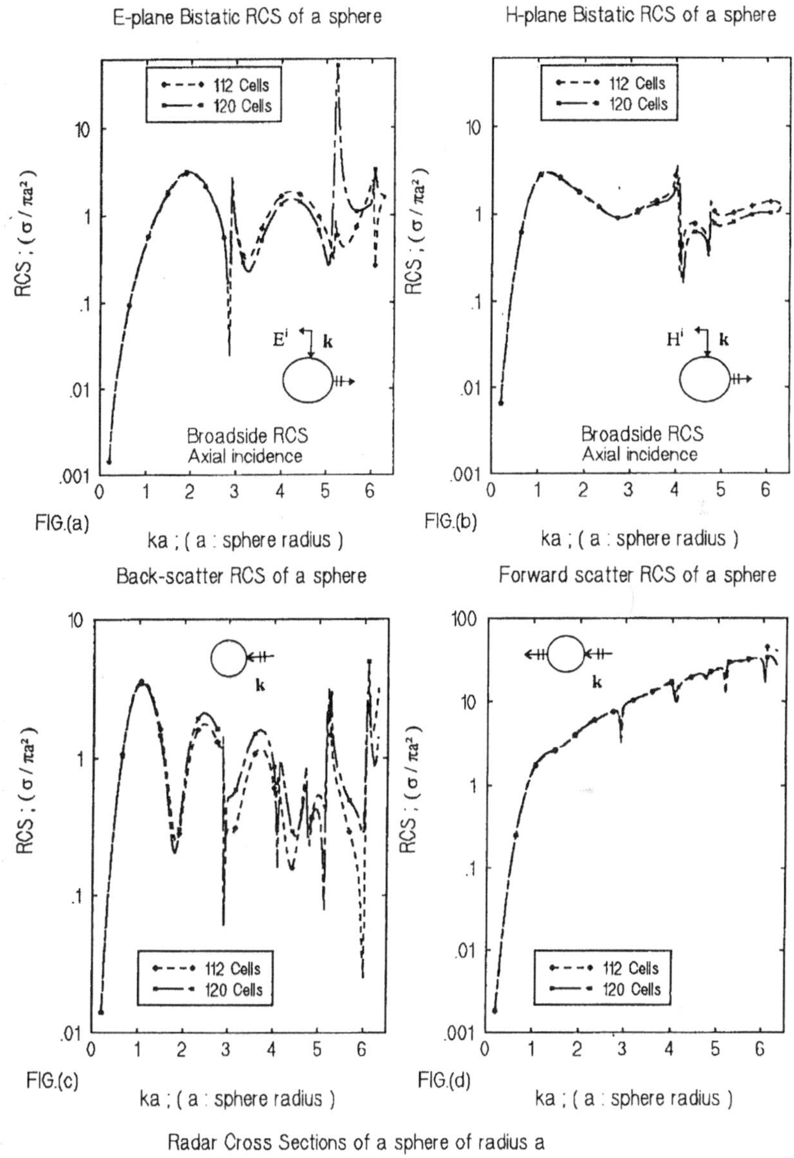

Figure 12.7. *Study of the convergence of the digital procedure*

1) Parasite models:

We can observe that the MFIE does not give a single solution when the frequency of the incident plane wave is in the vicinity of the conducting sphere's resonance frequencies. For a sphere of radius a, the phenomenon of resonance is produced at ka ≈ 2.85, 4.02, 4.70, 5.16 and 6.04. This depends on the polarization of the light radiation field.

2) RCS models:

Figures 12.8 and 12.9 show that when the value ka increases (the sphere becomes large electrically), the force of the dispersion field toward the front increases in consequence.

Studies of the diffusion by a sphere with radius a, illuminated by an incident field of the linearly polarized pane wave depending on the direction in space, are shown in Figure 12.10. These include the bistatic lines RCS 2D of the E- and H-plane and the bistatic RCS 3D diagrams coparellized to the different frequencies in the range of convergence. In the analysis, the radius is incident in the plane Φ = 90° along the axis (θi = 0°, Φi = 90°) and a converged discretization calculation is used. It is clear that the motifs E and H are the sections Φs = 90° of models σθθ and σΦΦ, respectively.

For any section θs of motifs σθθ and σΦΦ, the diffused field takes its crest value at Φs = 90° (plane of incidence). Because of the structure's axial symmetry, σθθ and σΦΦ, the schemas show a symmetry in relation to the plane Φs = 90° and in this plane the crossed polarizations (θΦ and Φθ) are null.

Calculations E and H show that the sphere concentrates the diffused power in the direction of back-scatter (the motifs are behind, i.e. σ is maximum in the direction θ = 0°) at ka = 1.046. On increasing the frequency of analysis (increasing ka), the height of the peak increases and the peak moves in a direct direction (the motifs become forward directors) and often, more peaks and related nulls appear.

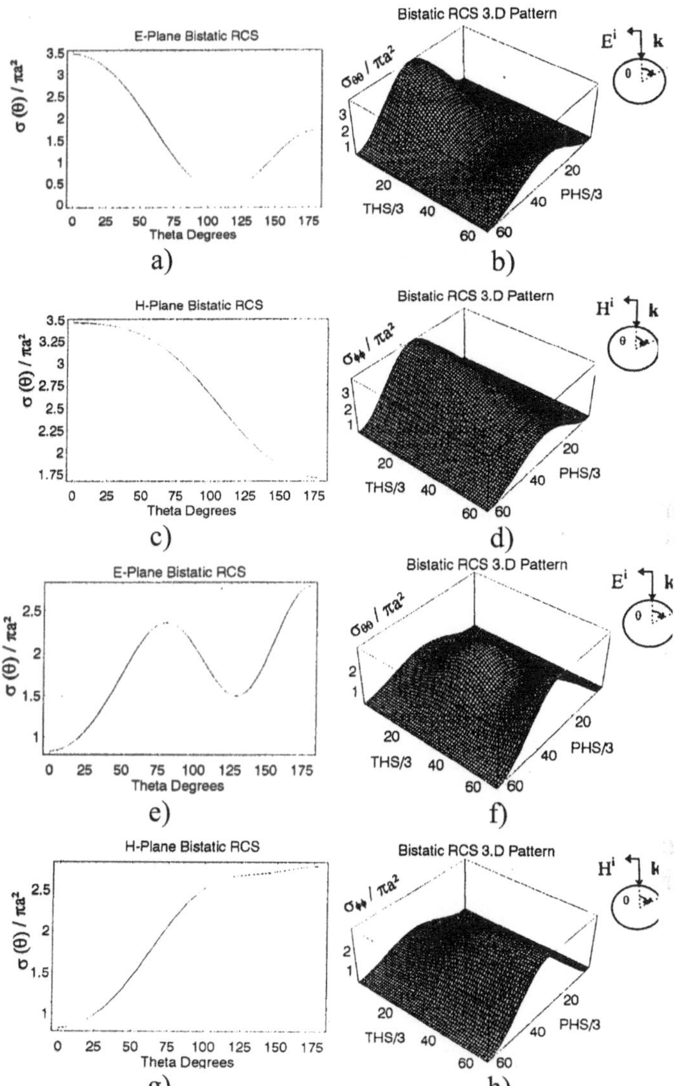

Figure 12.8. *a) E-plane bistatic RCS of the sphere for ka = 1.046; b) Bistatic RCS 3D pattern of the sphere for ka = 1.046; c) H-plane bistatic RCS of the sphere for ka = 1.046; d) Bistatic RCS 3D pattern H-plane of the sphere for ka = 1.046; e) E-plane E bistatic RCS of the sphere for ka = 1.57; f) H-plane bistatic RCS 3D pattern of the sphere for ka = 1.57; g) H-plane bistatic RCS of the sphere for ka = 1.57; h) RCS Bistatic 3D pattern of the sphere for k = 1.57*

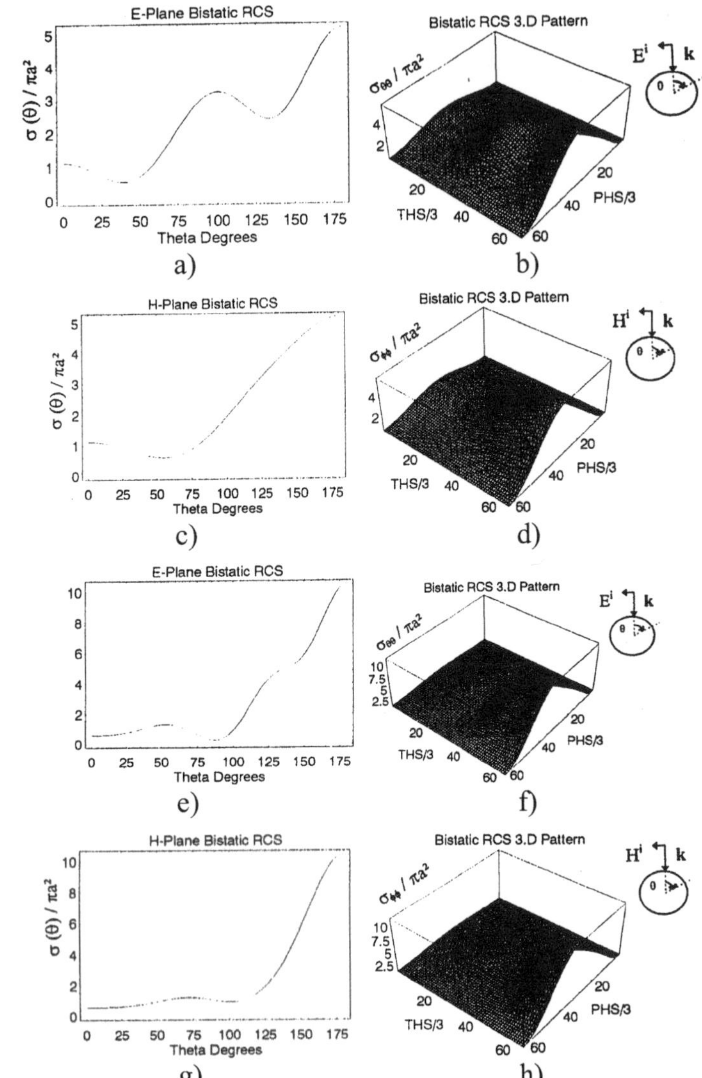

Figure 12.9. *a) E-plane bistatic RCS of the sphere for k = 2.092; b) RCS Bistatic 3D pattern of the sphere for k = 2.092; c) H-plane bistatic RCS of the sphere for k = 2.092; d) Bistatic RCS 3D pattern of the sphere for k = 2.092; e) E-plane bistatic ERCS of the sphere for k = 3.14; f) E-plane bistatic RCS 3 D pattern of the sphere for k = 3.14; g) H-plane bistatic RCS of the sphere for k = 3.14; h) Bistatic RCS 3 D pattern of the sphere for k = 3.14*

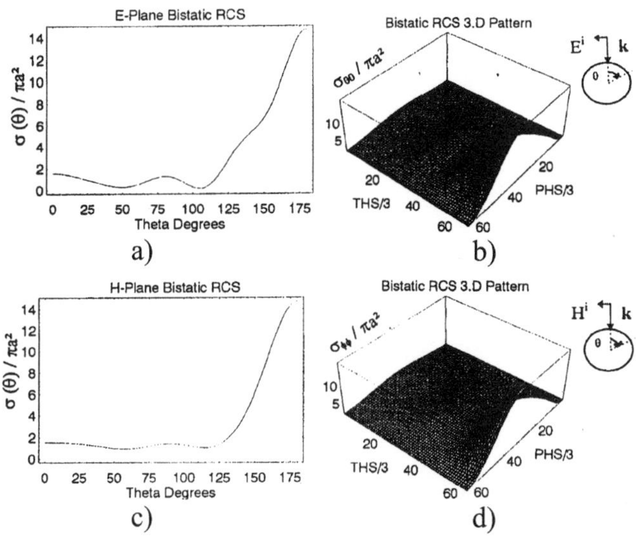

Figure 12.10. *a) E-plane bistatic RCS of the sphere for ka = 3.78; b) Bistatic RCS 3D pattern of the sphere for ka = 3.78; c) H-plane bistatic RCS of the sphere for ka = 3.78; d) Bistatic RCS 3D pattern of the sphere for ka = 3.78*

12.3.6. *Conclusion*

Compressed sensing is a broad field of application in the study of scattering in electromagnetism. We have tried to give an application to scattering in a sphere illuminated by an electromagnetic field. It emerges from the study that in terms of convergences, compressed sensing applied with the method of moments gives very promising results.

Conclusion

Compressed sensing in LiFi and WiFi systems has opened the way for a vast field for investigation. The combination of research in mathematics and information theory has laid the groundwork for new perspectives in signal processing and electromagnetism. What was thought initially to be the resolution of an inverse problem has revolutionized domains such as signal compression and acquisition. The constraint formed by the sampling rate has been completely removed by concepts such as subsampling and oversampling. Research published in the domain of WiFi networks promises future developments. We can legitimately consider that the acquisition of a complete signal is no longer necessary. A sliver of signal can reconstruct an entire panoply.

In wireless networks, several advances have been made: in particular we cite constraints in the design of wireless networks between the degree of compression reached by the algorithm and the necessary power imposed by the algorithm. A synergy between coding and the algorithm's communication aspects is vital. It is expected that data compression will increase in the next decade and have greater resolution in the deployment of WiFi networks.

In the domain of wavelets, the advent of compressed sensing has also brought about considerable change. Attempts to unify approaches in compressed sensing in signal treatment have been made and we cite the combination of random sampling and compressed sensing that has brought about a non-negligible rate of innovation. Classic optimization methods using the gradient have also undergone a notable evolution through the application of compressed sensing in the choice of norms for regularization or reconstruction. In the domain of target radar pursuit, compressed sensing has also brought about its share of change.

In wireless networks, promising elements for routing packets and the consumption of power have also been demonstrated.

Bibliography

[AFF 14] AFFELDT R., HAGIWARA M., SENIZERGUES J., "Formalization of Shannon's theorems", *Journal of Automated Reasoning*, vol. 53, pp. 63–103, 2014.

[AGA 07] AGARWAL V., GRIBOK A.V., ABIDI M.A., "Image restoration using L1 norm penalty function", *Inverse Problems in Science and Engineering*, vol. 15, no. 8, pp. 785–809, 2007.

[AHL 06] AHLUND C., BRANNSTROM R., ANDERSSON K. *et al.*, "Port-based multihomed mobile IPv6 for heterogeneous networks", *Local Computer Networks*, pp. 567–568, 2006.

[AI 15] AI X.F., LUO Y.J., ZHAO G.Q., "Transient interference excision and signal recovery in other using robust principal component", *Electronics Letters*, vol. 51, no. 17, pp. 1363–1364, 2015.

[AIJ 13] AIJAZ A., AGHVAMI H., AMANI M., "A survey on mobile data offloading: technical and business perspectives", *IEEE Wireless Communications*, vol. 20, pp. 104–112, 2013.

[AIS 08] AISSAOUI A., HAMMOUDI Z., FARROUKI A. *et al.*, "Adaptive pseudo-noise code acquisition scheme using automatic censoring for DS/SS communication in frequency-selective Rayleigh fading channel", *IET Communications*, vol. 2, no. 2, pp. 359–365, 2008.

[ALJ 05] ALJUNID S.A., SAMA M.D.A., OTHMAN M. *et al.*, "Development of modified double-weight code and its implementation in multi-rate transmissions networks communication", *13th IEEE International Conference*, vol. 1, pp. 1531–2216, 2005.

[ALJ 12] ALJUNID S.A., ABD T.H., FADHIL H.A., "New approach for evaluation of the performance of spectral amplitude coding-optical code division multiple access system on high-speed data rate", *IET Communications*, vol. 6, no. 12, pp. 1742–1749, 2012.

[AND 07] ANDRECUT M., ESTE R.A., KAUFFMAN S.A., "Competitive optimization of compressed sensing", *Journal of Physics A: Mathematical and Theoretical*, vol. 40, p. 305, 2007.

[ANU 06] ANUAR M.S., SAAD N.M., MOHAMMED A. *et al.*, "Development of a zero cross correlation code for spectral amplitudecodingoptical code division multiple access (OCDMA)", *IJCSNS International Journal of Computer Science and Network Security*, vol. 6, no. 12, 2006.

[ANU 07] ANUAR M.S., ALJUNID S.A., BADLISHAH R. *et al.*, "Performance analysis of optical zero cross correlation in OCDMA system", *Journal of Applied Sciences*, vol. 7, no. 23, pp. 3819–3822, 2007.

[ANU 09] ANUAR M.S., ALJUNID S.A., SAAD N.M. *et al.*, "New design of spectral amplitude coding in OCDMA with zero cross-correlation", *Optics Communications*, vol. 282, pp. 2659–2664, 2009.

[ASY 17] ASYHARI A.T., BRINK S., "Orthogonal or superimposed pilots? a rate-efficient channel estimation strategy for stationary mimo fading channels", *IEEE Transactions on Wireless Communications*, vol. 16, no. 5, pp. 2776–2789, 2017.

[AWA 07] AWAL M.A., DAS D.K., KANCHANASUT K., "Performance behavior of MANET with a fixed long delay satellite link", Technical report no. TR_2007-3, Terlab, AIT, Thailand, 2007.

[AZA 16] AZAM M.F.I., NI Q., ANSARI E.A., "Intelligent energy efficient localization using variable range beacons in industrial wireless sensor networks", *IEEE Transactions on Industrial Informatics*, vol. 12, no. 6, pp. 2206–2216, 2016.

[BAC 12] BACH F., JENATTON R., MURIEL J. *et al.*, "Structured sparsity through convex optimization", *Statistical Science*, vol. 27, no. 4, pp. 450–468, 2012.

[BAG 10] BAGADI K.P., DAS S., "MIMO-OFDM channel estimation using pilot carries", *International Journal of Computer Applications* (0975–8887), vol. 2, no. 3, pp. 81–88, 2010.

[BAJ 07] BAJWA W.U., HAUPT J.D., RAZ G.M. *et al.*, "Toeplitz structured compressed sensing matrices", *IEEE/Sp 14th Workshop on Statistical Signal Processing*, Proceedings August 26-29, Mona Terrace Convention Center Madison Wisconsin USA, p. 294 – 298, 2007.

[BAR 07] BARANIUK R.G., "Compressive sensing (Lecture notes)", *IEEE Signal Processing Magazine*, vol. 24, no. 4, pp. 118–121, 2007.

[BAR 10] BARANIUK R.G., CEVHER V., DUARTE M.F. *et al.*, "Model based compressive sensing", *IEEE Transactions on Information Theory*, vol. 56, no. 4, pp. 1982–2001, 2010.

[BEA 04] BEATY M.G., DODSON M.M., "The Whittaker-Kotel'nikov-Shannon theorem, spectral translates and Plancherel's formula", *The Journal of Fourier Analysis and Applications*, vol. 10, no. 2, pp. 179–199, 2004.

[BEE 95] DE BEEK J.-J.V., EDFORS O., SANDELL M. *et al.*, "On channel estimation in OFDM systems", *Proceedings of IEEE 45th Vehicular Technology Conference*, Chicago, IL, pp. 815–819, 1995.

[BEL 57] BELL D., *Information Theory and Its Engineering Applications*, Pitman and Sons, London, 1957.

[BEN 98] BENNET C.H., SCHOR P.W., "Quantum information theory", *IEEE Transactions on Information Sciences*, vol. 44, no. 6, pp. 2724–2742, 1998.

[BEN 12] BENKRINAH S. *et al.*, "An adaptive serial acquisition of PN sequence in nonhomogeneous AWGN channel using artificial neural networks", *IEEE International Conference on Complex Systems (ICCS)*, Morocco, 2012.

[BEN 13a] BENKRINAH S. *et al.*, "An adaptive serial acquisition scheme of PN sequence in nonhomogenous Rayleigh fading channel using artificial neural networks", *The International Conference on Electronics & Oil: From Theory to Applications (ICEO)*, Algeria, 2013.

[BEN 13b] BENKRINAH S., BENSLAMA M., "Effect of multi-users and multipaths on the performance of an adaptive serial acquisition scheme for DS/CDMA systems", *IEEE 8th International Workshop on Systems, Signal Processing and Their Applications (WoSSPA)*, Algeria, 2013.

[BEN 14] BENSLAMA M., BAITICHE M., MESSAI A., *Electromagnetic Scattering of Arbitrary Shaped Structures*, LAP Lambert Academic Publishing, Saarbrücken, 2014.

[BER 96] BERTSEKAS D., TSITSIKLIS J., *Neuro-Dynamic Programming*, Athena Scientific, Belmont, 1996.

[BER 14] BERBRA K. *et al.*, "PN code acquisition using smart antenna and adaptive thresholding CFAR based on ordered data variability for CDMA communications", *Progress in Electromagnetics Research B*, vol. 57, pp. 139–155, 2014.

[BIN 11] BIN W., YUMEI Z., YONGXU Z. *et al.*, "Design and implementation of an IEEE 802.11 baseband OFDM transceiver in 0.8 μm CMOs", *Journal of Semiconductors*, vol. 32, no. 5, p.0550011-0550015, 2011.

[BLU 09] BLUMENSATH T., DAVIES M.E., "Iterative hard thresholding for compressed sensing", *Applied and Computational Harmonic Analysis*, vol. 27, no. 3, pp. 265–274, 2009.

[BOC 09] BOCHE H., MÖNICH U.J., "Behavior of Shannon's sampling series for hardy spaces", *Journal of Fourier Analysis and Applications*, vol. 15, pp. 404–412, 2009.

[BOR 04] BORKAR V.S., MITTER S.K., VENKATESH S.R., "Variations on a theme by Neyman and Pearson", *Sankhya, The Indian Journal of Statistics*, vol. 66, no. 2, pp. 292–305, 2004.

[BOR 16] BORDOY J., TRAUB-ENS A., SADR A. *et al.*, "Bank of Kalman filters in closed loop for robust localization using unsynchronized beacons", *IEEE Sensors Journal*, vol. 16, no. 19, p. 7142, 2016.

[BRO 09] BROMBERG Y., KATZ O., SILBERBERG Y., "Ghost imaging with a simple detector", *Physical Review A*, vol. 79, p. 053840, 2009.

[BRU 99] BRUCKNER C., ZEILINGER A., "Operationaly invariant information in quantum mechanics", *Physical Review Letters*, vol. 83, no. 17, pp. 3354–3357, 1999.

[BRU 00] BRUCKNER C., ZEILINGER A., "Conceptual inadequacy of the Shannon information in quantum mechanics", *Physical Review A*, vol. 63, no. 2, pp. 3354–3357, 2000.

[BUD 02] BUDDHIKOT M., CHANDRANMENON G., HAN S. *et al.*, "Integration of 802.11 and third-generation wireless data networks", Infocom 2003, San Francisco, March 30–April 3, 2002.

[BUL 01] BULUSU N., HEIDMANN J., ESTRIN D., "Density adaptive algorithms for beacon placement in wireless sensor networks", *Proceedings of the 21st International Conference on Distributed Computing Systems*, pp. 489–498, 2001.

[BUR 12] BURKHOLDER R.J., O'DONNEL A.N., COBURN W.O. *et al.*, "Sparse basis expansions for compressive sensing of electromagnetic scattering patterns computed using physical optics", *2012 International Conference on Electromagnetics in Advanced Applications (ICEAA)*, IEEE, Cape Town, South Africa, p. 434, 2012.

[CAI 11] CAI T., TIANG T., "Limiting laws of coherence of random matrices with applications to testing covariance structure and construction of compressed sensing matrices", *The Annals of Statistics*, vol. 39, no. 3, pp. 1496–1525, 2011.

[CAN 05] CANDES E.J., TAO T., "Decoding by linear programming", *IEEE Transactions on Information Theory*, vol. 51, no. 12, pp. 4203–4215, 2005.

[CAN 06a] CANDES E.J., ROMBERG J., TAO T., "Robust uncertainty principles, exact signal reconstruction from highly incomplete frequency information", *IEEE Transactions on Information Theory*, vol. 52, no. 2, pp. 489–509, 2006.

[CAN 06b] CANDES E.J., TAO T., "Near optimal signal recovery from random projections, universal encoding strategies", *IEEE Transactions on Information Theory*, vol. 52, no. 12, pp. 5406–5425, 2006.

[CAN 07] CANDES E.J., ROMBERG J., "Sparsity and incoherence in compressive sensing", *Inverse Problems*, vol. 23, no. 3, p. 969–985, 2007.

[CAN 08a] CANDES E.J., "The restricted isometry property and its implications for compressed sensing", *Comptes Rendus Mathematique*, vol. 346, pp. 589–592, 2008.

[CAN 08b] CANDES E.J., WAKIN M.B., "An introduction to compressive sampling", *IEEE Signal Processing Magazine*, vol. 25, no. 2, pp. 21–30, 2008.

[CAN 09] CANDES E.J., RECHT B., "Exact matrix completion via convex optimization", *Foundations of Computational Mathematics*, vol. 9, p. 717, 2009.

[CAO 12] CAO X.Y., CHEN M.S., WU X.L., "Application of compressed sensing to solving 3D electromagnetic scattering problems over a wide angle", *IEEE International Conference on Microwaves and Millimeter Wave Technology (ICMMT)*, 2012.

[CAV 02] CAVES C.M., FUCHS C.A., SCHACK R., "Unknown quantum states, The quantum de Finetti representation", *Journal of Mathematical Physics*, vol. 43, pp. 4537–4559, 2002.

[CHA 05] CHAN T.F., ESEDOGLU S., "Aspects of total variation regularized L1 function approximation", *SIAM Journal of Applied Mathematics*, vol. 65, no. 5, pp. 1817–1837, 2005.

[CHA 15] CHAUDHURI R.R., DUTTA K., SAHA A., "Review on LiFi Technology", *Advances in Optical Science and Engineering*, Springer Proceedings in Physics, vol. 166, pp. 479–485, 2015.

[CHE 00] CHEN C., ZEBKER H., "Network approaches to two-dimensional phase unwrapping: intractability and two new algorithms", *Journal of the Optical Society of America*, vol. 17, no. 3, pp. 401–414, 2000.

[CHE 01] CHEN S.S., DONOHO D.L., SAUNDERS M.A., "Atomic decomposition by basis pursuit", *SIAM Review*, vol. 43, no. 1, pp. 129–159, 2001.

[CHE 06] CHEIKH K., SOLTANI F., "Application of neural networks to radar signal detection in K-distributed clutter", *IEE Proceedings Radar, Sonar and Navigation*, vol. 153, no. 5, pp. 460–466, 2006.

[CHE 12] CHEN W., WASSEL I.J., "Energy efficient signal acquisition in wireless sensor networks: a compressed sensing framework", *IET Wireless Sensor Systems*, vol. 2, no. 1, pp. 1–8, 2012.

[COH 09] COHEN A., DAHMEN W., DEVORE R., "Compressed sensing and best k-term approximation", *Journal of the American Mathematical Society*, vol. 22, pp. 211–231, 2009.

[COL 02] COLERI S., ERGEN M., PURI A. *et al.*, "A study of channel estimation in OFDM systems", *IEEE 56th Vehicular Technology Conference*, September 2002.

[COS 98] COSTANTINI M., "A novel phase unwrapping method based on network programing", *IEEE Transactions on Geoscience and Remote Sensing*, vol. 36, no. 3, pp. 813–821, 1998.

[COS 99] COSTANTINI M., FARINA A., "A fast phase unwrapping algorithm for SAR interferometry", *IEEE Transactions on Geoscience and Remote Sensing*, vol. 37, no. 1, p. 452–460, 1999.

[DAV 73] DAVIES J.B., "A least-squares boundary residual method for the numerical solution of scattering problems", *IEEE Transactions on Microwave Theory and Techniques*, vol. 21, no. 2, pp. 99–104, 1973.

[DEL 09] DELESTRE F., SUN Y., "Pilot aided channel estimation for MIMO-OFDM systems", *London Communications Symposium*, 2009.

[DEL 13] DE LAMARE R.C., "Adaptive and interative multibranch MMSE decision feedback detection algorithms for multi antenna systems", *IEEE Transactions on Wireless Communications*, vol. 12, no. 10, pp. 5294–5308, 2013.

[DEU 00] DEUTSCH D., HAYDEN P., "Information flow in entangled quantum system", *Proceedings of the Royal Society A, Mathematical, Physical and Engineering Sciences*, vol. 456, pp. 1759–1774, 2000.

[DIA 02] DIAS J., LEITAO J., "The ZπM algorithm for interferometric image reconstruction in SAR/SAS", *IEEE Transactions on Image Processing*, vol. 11, no. 4, pp. 408–422, 2002.

[DIA 07] DIAS J., VALADAO G., "Phase unwrapping via graph cut", *IEEE Transactions on Image Processing*, vol. 16, no. 3, pp. 698–709, 2007.

[DJE 16] DJEFFAL N., BENSLAMA M., "Quantum key distribution in WDM router applications for secured data transmission", *Optical and Quantum Electronics*, vol. 48, p.68-75, 2016.

[DO 12] DO T.T., GAN L., NGUYEN N.H. *et al.*, "Fast and efficient compressive sensing using structurally random matrices", *IEEE Transactions on Signal, Processing*, vol. 60, no. 1, pp. 139–154, 2012.

[DON 89] DONOHO D.L., STARK B., "Uncertainty principles and signal recovery", *SIAM J. Ap. Math*, vol. 49, no. 3, pp. 906–931, 1989.

[DON 06a] DONOHO D.L., ELAD M., TELYAKOV V.N., "Stable recovery of sparse overcomplete representations in the presence of noise", *IEEE Transactions on Information theory*, vol. 52, no. 1, pp. 6–18, 2006.

[DON 06b] DONOHO D.L., "Compressed sensing", *IEEE Transactions on Information Theory*, vol. 52, no. 4, pp. 1289–1306, 2006.

[DON 06c] DONOHO D.L., "For most large underdetermined systems of linear equations the minimal l-norm solution is also the sparsest solution", *Communications on Pure and Applied Mathematics*, vol. 59, no. 6, pp. 797–829, 2006.

[DOR 12] DORWARD R.M., CALLAN P.A., "Economic considerations for the electronic control of next generation multi-channel optical transceiver for the access network", *14th International Conference on Transparent Optical Networks (ICTON)*, pp. 1–5, 2012.

[DOU 16] DOUDOU F.Z., DJEBBARI A., "Particle swarm optimization aided serial acquisition in distributed OSCFAR and CMLD for DS-CDMA systems in fading channels", *Wireless Personal Communication*, vol. 94, pp. 1–20, 2016.

[DOW 03] DOWLER A., NIX A., "Performance evaluation of channel estimation techniques in a multiple antenna OFDM system", *IEEE Vehicular Technology Conference*, vol. 1, pp. 1–4, 2003.

[DUA 05] DUARTE M.F., SARVOTHAM S., BARON D. *et al.*, "Distributed compressed sensing of jointly sparse signals", *IEEE Conference Record of the Thirty Ninth Asilomar Conference on Signals, Systems and Computers*, pp. 1537–1541, 2005.

[DUA 11] DUARTE M.F., ELDAR Y.C., "Structured compressed sensing: from theory to applications", *IEEE Transactions on Signal Processing*, vol. 59, no. 9, pp. 4053–4085, 2011.

[DUA 12] DUARTE M.F., SHEN G., ORTEGA A., "Signal compression in wireless sensor networks", *Philosophical Transactions Mathematical, Physical and Engineering Sciences*, vol. 370, no. 1958, pp. 118–135, 2012.

[DUC 08] DUCHI J., SHALEV-SCHWARTZ S., SINGER Y. *et al.*, "Efficient projections onto the l1 ball for learning in high dimensions", *Proceedings of the 25th Conference on Machine Learning* (ICML'08), pp. 272–279, 2008.

[EGI 04] EGIDI N., MAPONI P., "A comparative study of two fast phase unwrapping algorithms", *Applied Mathematics and Computation*, vol. 148, pp. 599–629, 2004.

[EKS 09] EKSTROM H., "Qos control in the 3GPP evolved packet system", *IEEE Communications Magazine*, vol. 47, no. 2, pp. 76–83, 2009.

[ELA 07] ELAD M., "Optimized projections for compressed sensing", *IEEE Transactions on Signal Processing*, vol. 55, no. 12, pp. 5695–5702, 2007.

[ELD 09] ELDAR Y.C., MISHALI M., "Robust recovery of signals from a structured union of subspaces", *IEEE Transactions on Information Theory*, vol. 55, no. 11, pp. 5302–5316, 2009.

[ELD 10] ELDAR Y.C., KUPINGER P., BOLCSKEI H., "Block sparse signals, uncertainty relations and efficient recovery", *IEEE Transactions on Signal Processing*, vol. 58, no. 6, pp. 3042–3054, 2010.

[ELG 15] ELGALA H., LITTLE T.D.C., "Polar-based OFDM and SC-FDE links toward energy-efficient Gbps transmission under IM-DD optical system constraints", *Journal of Optical Communications and Networking*, vol. 7, no. 2, pp. 277–284, 2015.

[ERG 15] ERGUL O., DINC E., AKAN O.B., "Communicate to illuminate, state of the art and research challenges for visible light communications", *Physical Communications*, vol. 17, pp. 72–85, 2015.

[ETS 04] ETSI EN 300 744 v1.5.1, Digital Video Broadcasting (DVB); Framing structure, channel coding and modulation for digital terrestrial television, November 2004.

[EVA 05] EVANS B., "Integration of satellite and terrestrial systems in future multimedia communications," *IEEE Wireless Communication*, vol. 12, no. 5, p.72-80, 2005.

[FAN 55] FAN K., HOFFMAN A.J., "Somme metric inequalities in the space of matrices", *Proceedings of the American mathematical Society*, vol. 6, no. 1, pp. 111–116, 1955.

[FAZ 03] FAZEL K., KAISER S., *Multicarrier and Spread Spectrum Systems*, John Wiley & Sons, New York, 2003.

[FAZ 11] FAZEL F., FAZEL M., STOJANOVIC M., "Random access compressed sensing for energy efficient underwater sensor networks", *IEEE Journal of Selected Areas in Communications*, vol. 29, no. 8, p. 1660, 2011.

[FIG 07] FIGUEIREDO M.A.T., NOWAK R.D., WRIGHT S.J., "Gradient projection for sparse reconstruction, application to compressed sensing and other inverse problems", *IEEE Journal of Selected Topics in Signal Processing*, vol. 1, no. 4, pp. 586–597, 2007.

[FIS 25] FISHER R., "Theory of statistical estimation", *Proceedings of The Cambridge Philosophical Society*, vol. 22, pp. 700–725, 1925.

[FLA 12] FLAMMIA S.T., GROSS D., LIU Y.K. *et al.*, "Quantum tomography via compressed sensing, error bounds, sample complexity and efficient estimators", *New Journal of Physics*, vol. 14, p. 095022, 2012.

[FLY 97] FLYNN T., "Two-dimensional phase unwrapping with minimum weighted discontinuity", *Journal of the Optical Society of America A*, vol. 14, no. 10, pp. 2692–2701, 1997.

[FOR 11] FORNASSIER M., RAHUT R., "Compressive sensing", *Handbook of Mathematical Methods in Imaging*, pp. 187–288, Springer, Berlin, 2011.

[FOZ 05] FOZUNBAL M., MCLAUGHLIN S.W., SCHAFER R.W., "On space-time-frequency coding over MIMO-OFDM systems", *IEEE Transactions on Wireless Communications*, vol. 4, no. 1, pp. 320–331, 2005.

[FRI 77] FRIED D., "Least-squares fitting a wave-front distortion estimate to an array of phase-difference measurements", *Journal of the Optical Society of America*, vol. 67, no. 3, pp. 370–375, 1977.

[FRI 88] FRIEDMAN J.H., TUKEY J.W., "A projection pursuit algorithm for explatory data analysis", in CLEVELAND W.S. (ed.), *The Collected Works of J.W. Tukey*, Chapman & Hall, New York, 1988.

[FUC 04] FUCHS J.J., "On sparse representations in arbitrary redundant bases", *IEEE Transactions on Information Theory*, vol. 50, no. 6, pp. 1341–1344, 2004.

[FUC 09] FUCHS J.J., "Des moindres carrés aux moindres déviations", 2009 GRETSI Symposium on the Treatment of Sigma and Images, available at https:// hal.inria.fr/inria-00504534, 2009.

[GAB 88] GABRIEL A.K., GOLDSTEIN R.M., "Crossed orbit interferometry: theory and experimental results from SIR-B", *International Journal of Remote Sensing*, vol. 9, pp. 857–872, 1988.

[GAR 11] GARDNER P.S., The serval project: practical wireless ad-hoc mobile telecommunications, available at: http:/github.com//serval project.org/, 2011.

[GIL 02] GILBERT A.C., GUHA S., INDYK P. *et al.*, "Near optimal sparse Fourier representations via sampling STOC'02", *Proceedings of the 34th Annual Symposium on Theory Computing*, pp. 152–161, 2002.

[GIL 10] GILLIARD R., "The Lifi lamp high efficiency high brightness light emitting plasma with long life and excellent color quality", *IEEE International Conference on Plasma Science*, 2010.

[GIL 11] GILLIARD R., DEVINCIS M., HAFIDI A. *et al.*, "Operation of the LiFi emitting plasma in resonant cavity", *IEEE Transactions on Plasma Science*, vol. 39, no. 4, pp. 1026–1033, 2011.

[GIR 00] GIRARD A., ROSENBERG C., KHEMIRI M., "Fairness and aggregation: a primal decomposition study, networking 2000", in PUJOLLE G. *et al.* (eds.), *Lecture Notes in Computer Science 1815*, Springer-Verlag, Heidelberg, 2000.

[GHI 94] GHIGLIA D.C., ROMERO L., "Robust two-dimensional weighted and unweighted phase unwrapping that uses fast transforms and iterative methods", *Journal of the Optical Society of America A*, vol. 11, pp. 107–117, 1994.

[GHI 98] GHIGLIA D.C., PRITT M.D., *Two Dimensional Phase Unwrapping: Theory, Algorithms and Software*, John Wiley & Sons, New York, 1998.

[GLI 97] GLISIC S., VUCETIC B., *Spread Spectrum CDMA Systems for Wireless Communications*, Artech House, Boston, 1997.

[GRI 12] GRIBONVAL R., CEVHER V., DAVIES M.E., "Compressible distributions for high dimensional statistics", *IEEE Transactions on Information Theory*, vol. 58, no. 8, pp. 5016–5034, 2012.

[HAA 99] HAAS Z.J. *et al.* (eds.), *IEEE Journal on Selected Areas in Communications – Special Issue on Wireless Ad Hoc Networks*, vol. 17, no. 8, 1999.

[HAC 12a] HACINI L., Schémas Hybrides d'Acquisition Adaptative des Codes PN pour des Communications DS-CDMA dans un Canal Rayleigh, PhD Thesis, Mentouri Constantine University, 2012.

[HAC 12b] HACINI L. *et al.*, "Adaptive hybrid acquisition of PN sequences based on automatic multipath cancellation in frequency-selective Rayleigh fading channels", *Wireless Personal Communication*, vol. 63, no. 1, pp. 147–66, 2012.

[HAF 05] HÄFFNER H., HANSEL W., ROOS C.F. *et al.*, "Scalable multiparticle entanglement of trapped ions", *Nature*, vol. 438, pp. 643–646, 2005.

[HAN 12] HAN B., HUI P., KUMAR V.S.A. *et al.*, "Mobile data offloading through opportunistic communications and social participation," *IEEE Transactions on Mobile Computing*, vol. 11, pp. 821–834, 2012.

[HAS 83] HASKELL B., "High Definition Television (HDTV)–Compatibility and Distribution", *IEEE Transactions on Communications*, vol. 31, no. 12, p.1308–1317, 1983.

[HAS 12] HASSANIEH H., INDYK P., KATABI D. *et al.*, "Simple and practical algorithm for sparse Fourier transform", *Proceedings of the 23rd Annual ACM-SIAM Symposium on Discrete Algorithms*, pp. 1183–1194, 2012.

[HAS 14] HASSANIEH H., SHI L., ABARI O. *et al.*, "GHz-wide sensing and decoding using the sparse Fourier transform", *INFOCOM, Proceedings IEEE*, 2014.

[HAU 08] HAUPT J., BAJWA W.U., RABBAT M. *et al.*, "Compressed sensing for networked data", *IEEE Signal Processing Magazine*, vol. 25, no. 2, pp. 92–101, 2008.

[HEI 49] HEISENBERG W., *The Principles of the Quantum Theory*, Dover Publications USA, 1949.

[HER 09] HERMAN M.A., STROHMER T., "High resolution radar via compressed sensing", *IEEE Transactions on Signal Processing*, vol. 57, no. 6, pp. 2275–2284, 2009.

[HER 14] HERMANS F., MCNANAMAR L., VOIGT T. *et al.*, "Supporting heterogenous LCD/camera links", *Proceedings of the 13th International Symposium on Information Processing in Sensor Networks*, pp. 289–290, 2014.

[HIL 09] FADHIL A., ALJUNID S.A., AHMAD R.B., "Design and demonstration of a novel spectral amplitude coding OCDMA code for suppression phase intensity induced noise", *Journal of Communications*, vol. 4, no. 1, pp. 3–7, 2009.

[HOL 73] HOLEVO A.S., "Bounds for the quantity of information transmitted by a quantum communication channel", *Problems of Information Transmission*, vol. 9, no. 3, pp. 177–183, 1973.

[HOW 13] HOWLAND G.A., HOWELL J.C., "Efficient high-dimensional entanglement imaging with a compressive-sensing double-pixel camera", *Physical Review X*, vol. 3, no. 1, p. 0110131-01101310, 2013.

[HOW 15] HOWARD J., Notes and images of the STASIA 906, available at: https://twitter.com/howardjp/status/14640240637, 2015.

[HOW 16] HOWLAND G.A., KNARR S.H., SCHNEELOCH J. *et al.*, "Compressively characterizing high-dimensional entangled states with complementary, random filtering", *Physical Review X*, vol. 6, pp.1–13, 2016.

[HUD 77] HUDGIN R., "Wave-front reconstruction for compensated imaging", *Journal of the Optical Society of America*, vol. 67, no. 3, pp. 375–378, 1977.

[HUN 98] HUNG A., MONTPETIT M., KESIDIS G., "ATM via satellite: a framework and implementation", *Wireless Networks – Special Issue: Hybrid and Satellite Communication Networks*, vol. 4, no. 2, pp. 141–153, 1998.

[INT 03] INTEL, Wireless Hotspots and Network Interworking Initiative, White paper, Intel, Research and Development, 2003.

[IZU 13] IZUMIKAWA H., KATTO J., "Rocnet: spatial mobile data offload with user-behavior prediction through delay tolerant networks", *WCNC IEEE 2013*, Shanghai, China, 2013.

[KAB 09] KABASHIMA Y., WADAYAMA T., TANAKA T., "A typical reconstruction limit for compressed sensing based on lp-norm minimization", *Journal of Statistical Mechanics: Theory and Experiment*, vol. 2009, p. 1–11, 2009.

[KAL 02] KALYUZHNY A., KOVTONYUK A., "Basis of the Principal Informative Components for signal estimation problems", *IEEE International Conference on Acoustics, Speech, and Signal Processing (ICASSP)*, vol. 2, pp. II1657–II1660, 2002.

[KAN 16] KANG B., CHUNG C., KIM J., "Energy efficient beacon listening scheme for periodic vital signal monitoring", *Electronics Letters*, vol. 52, no. 18, pp. 1516–1518, 2016.

[KAR 13] KARLICEK R.F., "Future directions in LED Applications", in LASANCE C.J.M and POPPE A. (eds), *Thermal Management for LED Applications*, Springer, Berlin, 2013.

[KAT 02] KATZ M., "Code acquisition in advanced CDMA networks", *OULU 2002*, University of Oulu, 2002.

[KIM 97] KIM C.J. *et al.*, "Acquisition of PN code with adaptive threshold for DS/SS communications", *IEE Electronics Letters*, vol. 33, pp. 1352–1354, 1997.

[KIM 98a] KIM C.J. *et al.*, "Adaptive acquisition of PN sequences for DS/SS communications", *IEEE Transactions on Communications*, vol. 46, no. 8, pp. 993–996, 1998.

[KIM 98b] KIM C.J. *et al.*, "Adaptive hybrid acquisition of PN sequences for DS/SS communications", *IEE Electronics Letters*, vol. 34, pp. 939–940, 1998.

[KIM 16] KIM G., LEE S., PARK H. *et al.*, "A request based adaptive beacon rate control scheme for vehicular Adhoc networks", *IEEE 8th International Conference on Ubiquitous and Future Networks (ICUFN)*, 2016.

[KHI 57] KHINCHIN A., *Mathematical Foundations of Information Theory*, Dover, New York, 1957.

[KNU 16] KNUDSON K., SAAB R., WARD R.L., "One-bit compressive sensing with norm estimation", *IEEE Transactions on Information Theory*, vol. 62, no. 5, pp. 2748–2758, 2016.

[KOL 65] KOLMOGOROV A., "Three approaches to the quantitative definition of information", *Problems of information Transmission*, vol. 1, pp. 4–7, 1965.

[KOL 68] KOLMOGOROV A., "Logical basis for information theory and probability theory", *Transactions on Information Theory*, vol. 14, pp. 662–664, 1968.

[KOL 83] KOLMOGOROV A., "Combinatorial foundations on information theory and the calculus of probabilities", *Russian Mathematical Surveys*, vol. 38, pp. 29–40, 1983.

[KOU 02] KOUTSOGIANNIS G.S., SORAGHAN J.J., "Kernel principal component analysis (kpca) for the de-noising of communication signals", *11th Conference European Signal Processing*, vol. 2, pp. II–1167–II-1680, 2002.

[KRO 08] KROUMA H. *et al.*, "Performance analysis of an adaptive threshold hybrid double-dwell system with antenna diversity for acquisition in DS-CDMA systems", *International Journal Of electrical, Computer, Energetic, Electronic and Communication Technology*, vol. 4, pp. 376–385, 2008.

[KRZ 12] KRZAKALA F., MEZARD M., SAUSSET F. *et al.*, "Probabilistic reconstruction in compressed sensing, algorithms, phase diagrams, and threshold achieving matrices", *Journal of Statistical Mechanics: Theory and Experiment*, vol. 2012, pp. 1–50, 2012.

[KUM 94] KUMAR V., GRAMA A., GUPTA A. *et al.*, *Introduction to Parallel Computing, Design and Analysis Algorithms*, Addison Wesley, Boston, 1994.

[LAN 72] LANDAU L.D., LIFSHITZ E.M., *Mechanics and Electrodynamics Volume 1*, Pergamon Press, Oxford, 1972.

[LAN 91] LANDAUER R., "Information is physical", *Physics Today*, vol. 44, pp. 23–29, 1991.

[LAN 96] LANDAUER R., "The physical nature of information", *Physics Letters A*, vol. 217, pp. 188–193, 1996.

[LEE 04] LEE C.D., KWAK J.S., LEE J.H., "Low-rank pilot-symbol-aided channel estimation for MIMO-OFDM systems", *IEEE Vehicular Technology Conference*, vol. 1, pp. 469–473, 2004.

[LI 02] LI Y., WINTERS J.H., SOLLENBERGER N.R., "MIMO-OFDM for wireless communications: signal detection with enhanced channel estimation", *IEEE Transactions on Communications*, vol. 50, no. 9, pp. 1471–1477, 2002.

[LI 11a] LI G., HAO J., LI X. *et al.*, "The design and realization of a serial type LCD based on ARM9", *IEEE International Conference on Multimedia Technology*, pp. 3720–3722, 2011.

[LI 11b] LI Y., SU G., HUI P. *et al.*, "Multiple mobile data offloading through delay tolerant networks," *Proceedings of the 6th ACM Workshop on Challenged Networks*, pp. 43–48, 2011.

[LI 90] LI F.K., GOLDSTEIN R., "Studies of multibaseline spaceborne interferometric synthetic aperture radars", *IEEE Transactions on Geoscience and Remote Sensing*, vol. 28, no. 3, pp. 88–97, 1990.

[LIA 99] LIANG B., HAAS Z., "Predictive distance-based mobility management for PCS networks", *Proceedings of the Joint Conference of the IEEE Computer and Communications Societies (INFOCOM)*, March 1999.

[LIA 01] LIANG Y.L., LAJOS H., "Serial acquisition of DS-CDMA signals in multipath fading mobile channels", *IEEE Transactions on Vehicular Technology*, vol. 50, no. 2, pp. 617–628, 2001.

[LIN 91] LIN Q., VESEKY J.F., ZEBKER H.A., "Topography estimation with interferometric synthetic aperture radar using fringe detection", *Proceedings of IGARSS'91*, pp. 2173–2176, 1991.

[LIU 15] LIU X.Y., ZHU Y., KONG L. *et al.*, "CDC: compressive data collection for wireless sensor networks", *IEEE Transactions on Parallel and Distributed Systems*, vol. 26, no. 8, pp. 2188–2197, 2015.

[LIT 09] LITVIN M.V., "Formulation of the Shannon theorem for a discrete noisy channel", *Radiophysics and Quantum Electronics*, vol. 52, no. 10, pp. 750–757, 2009.

[LOM 16] LOMBARDI O., HOLIK F., VANNI L., "What is Shannon information", *Synthese*, vol. 193, no. 7, pp. 1983–2012, 2016.

[LYN 07] LYNCH J.P., "An overview of wireless structural health monitoring for civil structures", *Philosophical Transactions of the Royal Society A*, vol. 365, pp. 345–372, 2007.

[LUM 15] LUM D.J., KNARR S.H., HOWELL J.C., "Fast Hadamard transforms for compressive sensing of joint systems: measurement of a 3.2 million dimensional bi-photon probability distribution", *Optics Express*, vol. 23, no. 21, pp. 27636–27649, 2015.

[LUO 10] LUO J., LIU X., ROSENBERG C., "Does compressed sensing improve the throughput of wireless sensor networks", *Proceedings of the IEEE International Conference on Communications, ICC'10*, Cape Town, South Africa, 2010.

[LUS 07] LUSTIG M., DONOHO D.L., PAULY J.M., "Sparse MRI, the application of compressed sensing for rapid MR imaging", *Magnetic Resonance in Medicine*, vol. 58, no. 6, pp. 1182–1195, 2007.

[MAC 69] MACKAY D.M., *Information, Mechanism and Meaning*, MIT Press, Cambridge, MA, 1969.

[MAD 93] MADSEN S.N., ZEBKER H.A., MARTIN J., "Topographic mapping using radar interferometry: processing techniques", *IEEE Transactions on Geoscience and Remote Sensing*, vol. 31, no. 1, pp. 246–256, 1993.

[MAD 11] MADADI M., TATA M., "Shannon information in record data", *Metrika*, vol. 74, pp. 11–31, 2011.

[MAL 09] MALLAT S., *A Wavelet Tour for Signal Processing. The Sparse Way*, Academic Press, Cambridge, 2009.

[MAH 10a] MAHMOUD H.M., MOUSA A.S., SALEEM R., "Channel estimation based in comb-type pilots arrangement for OFDM system over time varying channel", *Journal of Networks*, vol. 5, no. 7, pp. 766–772, 2010.

[MAH 10b] MAHMUDIMANESH M., KHELIL A., SURI N., "Reordering for better compressibility, efficient spatial sampling in wireless sensor networks", *IEEE International Conference on Sensor Networks, Ubiquitous and Trustworthy Computing*, 2010.

[MAR 93] MARIC S.V., KOSTIC Z.I., TITLEBAUM E.L., "A new family of optical code sequences for use in spread-spectrum fiber-optic local area networks", *IEEE Transactions on Communications*, vol. 41, pp. 1217–1221, 1993.

[MAR 12] MARVASTI F., AMINI A., HADDADI F. *et al.*, "A unified approach to sparse signal processing", *Eurasip Journal of Advances in Signal Processing*, vol. 44, pp. 1–45, 2012.

[MAS 92] MASSONNET D., "Perspectives de l'interférométrie par radar", *Bulletin SFPT*, no. 127, pp. 3–15, 1992.

[MAS 93a] MASSONNET D., PERLANT F., RABAUTE T., "Précision et niveau d'artéfacts dans les calculs de modèles numériques de terrain par interférométrie", *De l'optique au radar, les Applications de SPOT et ERS*, Cepadues-Editions, Paris, 1993.

[MAS 93b] MASSONNET D., RABAUTE T., "Radar interferometry: limits and potentials", *IEEE Transactions on Geoscience and Remote Sensing*, vol. 31, no. 2, pp. 445–464, 1993.

[MAS 94] MASSONNET D., ADRAGNA F., ROSSI M., "CNES general purpose SAR correlator", *IEEE Transactions on Geoscience and Remote Sensing*, vol. 32, pp. 636–643, 1994.

[MAS 13] MASOUM A., MERATNIA N., HAVINGA P.J.M., "A distributed compressive sensing technique for data gathering in wireless sensor networks", *The 4th International conference on Emerging Ubiquitous Systems and Pervasive Networks*, 2013.

[MAS 15] MASSA A., ROCCA P., OLIVERI G., "Compressive sensing in electromagnetic: a review", *IEEE Antennas and Propagation Magazine*, vol. 15, no. 1, p. 224, 2015.

[MIA 15] MIAO Y., SUN Z., WANG N. *et al.*, "Comparison studies of MANET satellite and MANET cellular network integrations", *IEEE International Conference on Wireless Communications & Signal Processing (WCSP)*, 2015.

[MIC 12] MICHAEL C., Algorithms and mechanisms for system optimization, MONET project report, 2012.

[MIG 11] MIGLIORE M.D., PINCHERA D., "Compressed sensing in electromagnetic, theory, applications and perspectives", *Proceedings of the 5th European Conference on Antennas and Propagation*, pp. 1969–1973, 2011.

[MIG 14] MIGLIORE M.D., "A simple introduction to compressed sensing/sparse recovery with applications in antenna measurements", *IEEE Antennas and Propagation magazine*, vol. 56, no. 2, pp. 14–26, 2014.

[MIS 11] MISHALI M., ELDAR Y.C., ELRON A.J., "Xampling, signal acquisition and processing union of subspaces", *IEEE Transactions on Signal Processing*, vol. 59, no. 10, p. 4719, 2011.

[MIX 11] MIXON D.G., BAJWA W.U., CALDERBANK R., "Frame coherence and sparse signal processing", *IEEE International Symposium on Information Theory*, 2011.

[MOS 09] MOSTOFI Y., SEN P., "Compressive cooperative sensing and mapping in mobile networks", *American Control Conference*, p. 3397, 2009.

[MUQ 99] MUQUET B., DE COURVILLE M., "Blind and semi-blind channel identification methods using second order statistics for OFDM systems", *IEEE International Conference Acoustics, Speech, and Signal Processing*, vol. 5, pp. 2745–2748, 1999.

[NEC 04] NECKER M.C., STUBER G.L., "Totally blind channel estimation for OFDM on fast varying mobile radio channels", *IEEE Transactions on Wireless Communications*, vol. 3, no. 5, pp. 1514–1525, 2004.

[NES 94] NESTEROV Y., NEMIROVSKII A., *Interior Point Polynomial Algorithms in Convex Programming*, SIAM Society for Industrial and Applied Mathematics, Philadelphia, 1994.

[NGO 03] LE-NGOC V., LEUNG V., TAKATS P. *et al.*, "Interactive multimedia satellite access communications", *IEEE Communications Magazine*, vol. 41, no. 7, pp. 78–85, 2003.

[OCO 14] O'CONNOR S.M., LYNCH P., GILBERT A.C., "Compressed sensing embedded in an operational wireless sensor network to achieve energy efficiency in long term monitoring applications", *Smart Materials and Structures*, vol. 23, pp. 085014–085030, 2014.

[OGA 13] OGATA Y., "The Shannon-McMillan theorem for AFC* systems", *Letters in Mathematical Physics*, vol. 103, pp. 1367–1376, 2013.

[OKA 09] OKA A., LAMPE L., "A compressed sensing receiver for UWB impulse radio in bursty applications like wireless sensor networks", *Physical Communications*, vol. 2, pp. 248–264, 2009.

[OH 02] OH H.S. *et al.*, "Adaptive hybrid PN code acquisition with antenna diversity in DS-CDMA systems", *IEICE Transactions on Communications*, vol. E85-B, no. 4, pp. 716–722, 2002.

[OH 05] OH H.S., HAN D.S., "An adaptive double-dwell PN code acquisition system in DS-CDMA communications", *Signal Processing*, vol. 85, no. 12, pp. 2327–2337, 2005.

[OLI 10] OLIVEIRA A., SUN Z., MONIER M. *et al.*, "On optimizing hybrid ad-hoc and satellite networks the monet approach," *Proceedings of Future Network and Mobile Summit Conference*, 2010.

[OLI 13] OLIVERI G., ROCCA P., MASSA A., "Compressive sensing as applied to electromagnetic; advances, comparisons and applications", *7th European Conference on Antennas and Propagation*, p. 1523, 2013.

[PAN 11] PANDE V.S., "Compressed sensing and sensibility", *Proceedings of the National Academy of Sciences of the United States of America*, vol. 108, no. 36, pp. 14173–14174, 2011.

[PET 04] PETROPULU A., ZHANG R., LIN R., "Blind OFDM channel estimation through simple linear pre-coding", *IEEE Transactions on Wireless Communications*, vol. 3, no. 2, pp. 647–655, 2004.

[PLA 36] PLANCHEREL M., POLYA G., "Fonctions entières et intégrales de Fourier multiples", *Commentarii mathematici Helvetici*, vol. 9, no. 1, pp. 224–248, 1936.

[PAT 93] PATI Y.C., REZAIIFAR R., P.S. KRISHNAPRASAD P.S., "Orthogonal matching pursuit: recursive function approximation with applications to wavelet decomposition", *IEEE Conference Record of the 27th Asilomar Conference on Signals, Systems and Computers*, 1993.

[PLE 05] PLENIO M.B., VIRMANI S., An introduction to entanglement measures, ArXiv preprint-ph/0504163, 2005-arxiv.org, 2005.

[POC 99] LE POCHER H., LEUNG V.C.M., GILLIES D.W., "Real-time multimedia scheduling policies for end-to-end delay jitter and loss guarantees across ATM satellite systems", *IEEE Journal on Selected Areas in Communications*, vol. 17, no. 2, pp. 314–325, 1999.

[POL 03] POLITIS C., CHEW K.A., TAFAZOLLI R., "Multilayer mobility management for all-IP networks: pure SIP vs. hybrid SIP/mobile IP" *Vehicular Technology Conference*, vol. 4, pp. 2500–2504, 2003.

[PRA 89] PRATI C., ROCCA F., GUARNIERI A., "Effects of speckle and additive noise on the altimetric resolution of SAR surveys", *IEEE International Geoscience and Remote Sensing Symposium, IGARSS'89*, pp. 2469–2472, 1989.

[PRA 90] PRATI C., ROCCA F., GUARNIERI A., *et al.* "Seismic migration for SAR focusing: international application", *IEEE Transactions on Geoscience and Remote Sensing*, vol. 28, pp. 627–639, 1990.

[PRA 02] PRADHAM S.S., KUSUMA J., RAMCHANDRAN K., "Compression in a dense microsensor network", *IEEE Signal Processing Magazine*, vol. 19, no. 2, pp. 51–60, 2002.

[PRE 98] PRESKILL J., Lectures notes for Physics 229: Quantum Information and Computation, 1998.

[PRU 86] PRUCNAL P., SANTORO M., TING F. *et al.*, "Spread spectrum fiber-optic local area network using optical processing", *Lightwave Technology*, vol. LT-4, pp. 547–554, 1986.

[PUS 12] PUSEY M.F., BARRET J., RUDOLPH T., "On the reality of the quantum state", *Nature Physics*, vol. 8, pp. 475–478, 2012.

[QUE 89] QUEGAN S., "Interpolation and sampling in SAR images", *Proceedings of IGARSS'89*, pp. 612–616, 1989.

[QUE 09] QUER G., MASIERO R., MUNARETTO D., "On the interplay between routing and signal representation for compressed sensing in wireless sensors networks", Dsp.rice.edu/files/Cs/quer_ITA2009, 2009.

[RAE 99] RAE H., "Performance analysis of a double-dwell serial search technique for cellular CDMA networks in the case of multiple pilot signals", *IEEE Transactions on Vehicular Technology*, vol. 48, no. 6, pp. 1819–1830, 1999.

[RAO 11] RAO K.R., KIM D.N., WANG J.J., *Fast Fourier Transform, Algorithms and Applications*, Springer, Berlin, 2011.

[RAS 11] RASTEGIN A.E., "Bounds on Shannon distinguishability in terms of portioned measures", *Quantum Information Processing*, vol. 10, pp. 123–138, 2011.

[RAS 13] RASHIDI C.B.M., ALJUNID S.A., FADHIL H.A. *et al.*, "Encoder-decoder design for SAC-OCDMA using flexible cross correlation (FCC) code algorithms", *4th International Conference on Photonics*, pp. 355–359, 2013.

[RAU 08] RAUHUT H., SCHNASS K., VANDERGHEYNST P., "Compressed sensing and redundant dictionaries", *IEEE Transactions on Information Theory*, vol. 54, no. 5, pp. 2210–2219, 2008.

[RE 13] RE E., JAYOUSI S., MOROSI S. *et al.*, "Salice project: satellite assisted localization and communication systems for emergency services", *IEEE Aerospace and Electronic Systems Magazine*, vol. 28, pp. 2–13, 2013.

[REC 10] RECHT B., FAZEL M., PARRILO P.A., "Guaranteed minimum rank solutions of linear matrix equations via nuclear norm minimization", *SIAM Review*, vol. 52, no. 3, pp. 471–501, 2010.

[RIO 17] RIOFRIO C.A., GROSS D., FLAMMIA S.T. *et al.*, "Experimental quantum compressed sensing for a seven qubit system", *Nature Comm.*, vol. 8, 15305, 2017.

[RON 03] RONGA L.S., PECORELLA T., FANTACCI R. *et al.*, "Real-time QoS DiffServ gateway implementation for satellite IP networks", available at: https://www.resarchgate.net/ publication/2911497, 2003.

[RUD 06] RUDELSON M., VERSHYIN R., "Sparse reconstruction by convex relaxation, Fourier and Gaussian measurements", *IEEE Annual Conference on Information Sciences and Systems*, 2006.

[RUF 10] RUFO J., RABADAN J., DELGADO F. *et al.*, "Experimental evaluation of video transmission through LED illumination devices", *IEEE Transactions on Consumer Electronics*, vol. 56, no. 3, pp. 1411–1416, 2010.

[RUM 86] RUMELHART D.E. *et al.*, "Learning representations by back-propagation errors", *Natures*, vol. 323. no. 9, pp. 533–536, 1986.

[SIC 04] SICHITIU M.L., RAMADURAI V., "Localization of wireless sensor networks with a mobile beacon", *IEEE international Conference on Mobile Ad-Hoc and sensor Systems*, pp. 174–183, 2004.

[SIM 14] SIMMONS R.N., WINTUCKY E.G., "Multiband multi tone millimeter wave frequency synthesizer for atmospheric propagation studies", *IEEE International Symposium on Antenna and Propagation*, APSURSI, 2014.

[SHA 48] SHANNON C., "The mathematical theory of communication", *Bell System Technical Journal*, vol. 27, pp. 379–423, 1948.

[SHE 06] SHEN Y., MARTINEZ E., Channel estimation in OFDM systems, Free Scale Semiconductor, AN3059, 2006.

[SHI 01] SHIN O.S., LEE K.B., "Utilization of multipath for spread-spectrum code acquisition in frequency-selective Rayleigh fading channels", *IEEE Transactions on Communications*, vol. 49, no. 4, pp. 734–743, 2001.

[SHI 08a] SHI G., LIN J., CHEN X. *et al.*, "UWB echo signal detection with ultralow rate sampling based on compressed sensing", *IEEE Transactions on Circuits and systems, II: Express Briefs*, vol. 55, no. 4, pp. 379–383, 2008.

[SHI 08b] SHI X.J., ZHANG Y.H., "Quantitative comparison of phase unwrapping algorithms for SAR interferometry", *Microwave Conference, Asia-Pacific IEEE*, Macau, pp. 1–4, 2008.

[SHI 10] SHI W., ZHU Y., YAO Y., "Discussion about the DCT/FFT phase-unwrapping algorithm for interferometric applications", *Optik–International Journal for Light and Electron Optics*, vol. 121, no. 16, pp. 1443–1449, 2010.

[SHU 15] SHUYING L. *et al.*, "Hyperchaotic spread spectrum sequences selection and its application in DS-CDMA system", *Radioelectronics and Communications Systems*, vol. 58, no. 9, pp. 404–410, 2015.

[SOF 16] SOFWAN A., BARKAT M., AL. QAHTANI S.A., "PN code acquisition using smart antennas and adaptive thresholding for spread spectrum communications", *Wireless Network*, vol. 22, no. 1, pp. 223–234, 2016.

[STE 01] STEWART R., METZ C., "SCT new transport protocol for TCP/IP", *IEEE Internet Computing*, vol. 5, no. 6, pp. 64–69, 2001.

[SUN 14] SUNG W.T., CHEN J.H., LIN J.S. *et al.*, "Adaptive weighted data fusion of wisdom LED lighting control system design and research", *IEEE International Symposium on Computer, Consumer and Control*, pp. 426–429, 2014.

[SZA 91] SZAREK S.J., "Conditions numbers of random matrices", *Journal of Complexity*, vol. 7, pp. 131–149, 1991.

[TAK 11] TAKEDA H., KABASHIMA Y., "Statistical mechanical assessment of a reconstruction limit of compressed sensing, toward theoretical analysis of correlated signals", *EPL (Europhysics Letters)*, vol. 95, no. 1, p. 18006, 2011.

[TAK 13] TAKEDA H., KABASHIMA Y., "Reconstruction algorithm in compressed sensing based on maximum a posteriori estimation", *Journal of Physics: Conference Series*, vol. 473, p. 012003, 2013.

[TAL 11] TALEB T., AOUL Y.H., AHMED T., "Challenges, opportunities, and solutions for converged satellite and terrestrial networks", *IEEE Wireless Communications*, vol. 18, pp. 46–52, 2011.

[TAM 10] TAMBERG G., "On truncation errors of some generalized Shannon sampling operators", *Numerical Algorithms*, vol. 55, pp. 367–382, 2010.

[TIA 07] TIAN Z., GIANNAKIS G.B., "Compressed sensing for wideband cognitive radios", *IEEE International Conference on acoustic, speech and signal Processing, ICASSP*, 2007.

[TIB 96] TIBSHIRANI R., "Regression shrinkage and selection via the Lasso", *Journal of the Royal Statistical Society. Series B (Methodological)*, vol. 58, no. 1, pp. 267–288, 1996.

[TIM 13] TIMPSON C., *Quantum Information Theory and the Foundations of Quantum Mechanics*, Oxford University Press, Oxford, 2013.

[TON 14] TONOLINI F., CHAN S., AGNEW M. *et al.*, "Reconstructing high-dimensional two-photon entangled states via compressive sensing", *Scientific Reports*, vol. 4, pp. 1–5, 2014.

[TRO 07] TROPP J.A., GILBERT A.C., "Signal recovery from random measurements via orthogonal matching pursuit", *IEEE Transactions on Information Theory*, vol. 53, no. 12, pp. 4655–4666, 2007.

[TSA 02] TSANG L., KONG J.A., DING H., *Scattering of Electromagnetic Waves, Theories and Applications*, John Wiley and Sons, New York, 2002.

[TSA 06] TSAIG Y., DONOHO D.L., "Extensions of compressed sensing", *Signal Processing*, vol. 86, no. 3, pp. 549–571, 2006.

[TUR 37] TURING A.M., "On computable numbers, with an application to the entscheidungsproblem", *Proceedings of the London Mathematical Society*, vol. s2-42, no. 1, pp. 230–265, 1937.

[TUR 50] TURING A.M., "Computing machinery and intelligence", *Mind*, vol. LIX, pp. 433–460, 1950.

[UDD 13] UDDIN M., NADEEM T., "RF-Beep, a light ranging scheme for smart devices", *IEEE international Conference on Pervasive Computing and Communications*, 2013.

[UHL 00] UHLMANN A., "On partial fidelities", *Reports on Mathematical Physics*, vol. 45, no. 3, pp. 407–418, 2000.

[VUK 10] VUKADINOVIC V., KARLSSON G., "Spectrum and energy efficiency of mobility-assisted podcasting in cellular networks," *MobiOpp '10 Proceedings of the Second International Workshop on Mobile Opportunistic Networking*, pp. 51–57, 2010.

[WAN 13] WANG Q.Q., JIN T., ZHU H.B. *et al.*, "Analyzing electromagnetic scattering using characteristic basis function method with compressed sensing", *2013 6th International Congress on Image and Signal Processing*, p. 1633, 2013.

[WAN 14] WANG Y., HUANG H.F., WU M.Q., "A new phase unwrapping method for interferograms with discontinuities", *Proceedings of the IEEE Radar Conference,* 2014.

[WAT 89] WATKINS C.J., Learning from delayed rewards, PhD Thesis, Kings College, Cambridge, 1989.

[WEI 01] WEI Z., SHALABY H.M., GHAFOURI S.H., "New code families for fiber-brag grating-based spectral-amplitude-coding optical CDMA systems", *IEEE Photonics Technology Letters*, vol. 13, no. 8, pp. 890–892, 2001.

[WEI 02] WEI Z., GHAFOURI-SHIRAZ H., "Codes for Spectral-Amplitude Coding Optical CDMA Systems", *Journal of Lightwave Technology*, vol. 20, no. 8, pp. 1284–1291, 2002.

[WEN 01] WENG C.S. WU J., "Optical orthogonal codes with non-ideal cross- correlation", *Journal of Lightwave Technology*, vol. 19, pp. 1856–1863, 2001.

[WEY 50] WEYL H., *The Theory of Groups and Quantum Mechanics*, Dover Publications, 1950.

[WIG 14] WIGREN T., CENTONZA A., GUNNARSON F., Closeness positioning in wireless networks, WO20140021771A2, patent 2014.

[WIL 16] WILDE M.M., *Quantum Information Theory*, 2nd ed., Cambridge University Press, Cambridge, 2016.

[WON 03] WONG K.K., MURCH R.D., BEN LETAIEF K., "A joint channel diagonalization for multiuser MIMO antenna systems", *IEEE Transactions on Wireless Communications*, vol. 2, no. 4, pp. 773–786, 2003.

[XIA 11] XIANG L., LUO J., VASILAKOS A., "Compressed data aggregation for energy efficient wireless sensor networks*", Eighth Annual IEEE Communications Society Conference On Sensor, Mesh and Ad Hoc Communications and Networks*, 2011.

[XU 14] XU Y., KABASHIMA Y., ZDEBOROVA L., "Bayesian signal reconstruction for 1 bit compressed sensing", *Journal of Statistical Mechanics: Theory and Experiment*, vol. 2014, pp. 1–22, 2014.

[YAN 04] YANG C.D., LEI P.H., PONG D.J. *et al.*, "Edge coupled in GaAs Pin photodiode with a light funnel waveguide", *IEEE Journal of Quantum Electronics*, vol. 40, no. 11, pp. 1607–1613, 2004.

[YAN 11] YANG J., ZHANG Y., "Alternating direction algorithms for l1 problems in compressive sensing", *SIAM Journal on Scientific Computing*, vol. 33, no. 1, pp. 250–278, 2011.

[YAN 16] YANG J. *et al.*, "Spreading-sequence design for partially connected multirelay networks under multipath fading", *IEEE Transactions on Vehicular Technology*, vol. 65, no. 3, pp. 1420–1433, 2016.

[ZEB 94] ZEBKER H.A., WERNER C.L., ROSEN P.A. *et al.*, "Accuracy of topographic maps derived from ERS-1 Interferometric radar", *IEEE Transactions on Geoscience and Remote Sensing*, vol. 32, pp. 823–836, 1994.

[ZHA 12] ZHANG L., LUO J., GUO D., "Neighbor discovery for wireless networks via compressed sensing", arXiv:2012.1007v4 [csNI], 2012.

[ZHA 13] ZHAO Y., FENG J., ZHANG B.C. *et al.*, "Current progress in sparse signal processing applied to radar imaging", *Science China Technological Sciences*, vol. 56, no. 12, pp. 3049–3054, 2013.

[ZHA 14] ZHANG Y., WANG S., JI G. *et al.*, "An improved quality guided phase unwrapping method and its applications to MRI", *Progress in Electromagnetics research*, vol. 145, pp. 273–286, 2014.

[ZHO 02] ZHOU X., LIU N., BARAS J., Flow control at satellite gateways, Technical Report ISR TR 2002-37, Institute for Systems Research, University of Maryland, 2002.

[ZHU 11] ZHU J. WANG X., "Model and protocol for energy-efficient routing over mobile ad hoc networks", *IEEE Transactions on Mobile Computing*, vol. 10, pp. 1546–1557, 2011.

Index

I, L

M

N, O, P

Q, R

S

T, U, V, X

FSC
www.fsc.org
MIX
Papier | Fördert gute Waldnutzung
FSC® C083411

Kolibri 360 GmbH
Ferdinand-Jühlke-Straße 7
99095 Erfurt, Deutschland
produktsicherheit@kolibri360.de